Studies in Systems, Decision and Control

Volume 85

Series editor

Janusz Kacprzyk, Polish Academy of Sciences, Warsaw, Poland
e-mail: kacprzyk@ibspan.waw.pl

About this Series

The series "Studies in Systems, Decision and Control" (SSDC) covers both new developments and advances, as well as the state of the art, in the various areas of broadly perceived systems, decision making and control-quickly, up to date and with a high quality. The intent is to cover the theory, applications, and perspectives on the state of the art and future developments relevant to systems, decision making, control, complex processes and related areas, as embedded in the fields of engineering, computer science, physics, economics, social and life sciences, as well as the paradigms and methodologies behind them. The series contains monographs, textbooks, lecture notes and edited volumes in systems, decision making and control spanning the areas of Cyber-Physical Systems, Autonomous Systems, Sensor Networks, Control Systems, Energy Systems, Automotive Systems, Biological Systems, Vehicular Networking and Connected Vehicles, Aerospace Systems, Automation, Manufacturing, Smart Grids, Nonlinear Systems, Power Systems, Robotics, Social Systems, Economic Systems and other. Of particular value to both the contributors and the readership are the short publication timeframe and the world-wide distribution and exposure which enable both a wide and rapid dissemination of research output.

More information about this series at http://www.springer.com/series/13304

Vladimir V. Breer · Dmitry A. Novikov
Andrey D. Rogatkin

Mob Control: Models of Threshold Collective Behavior

 Springer

Vladimir V. Breer
Trapeznikov Institute of Control Science
Russian Academy of Sciences
Moscow
Russia

Andrey D. Rogatkin
Trapeznikov Institute of Control Science
Russian Academy of Sciences
Moscow
Russia

Dmitry A. Novikov
Trapeznikov Institute of Control Science
Russian Academy of Sciences
Moscow
Russia

ISSN 2198-4182 ISSN 2198-4190 (electronic)
Studies in Systems, Decision and Control
ISBN 978-3-319-84763-4 ISBN 978-3-319-51865-7 (eBook)
DOI 10.1007/978-3-319-51865-7

Printed on acid-free paper

This Springer imprint is published by Springer Nature
The registered company is Springer International Publishing AG
The registered company address is: Gewerbestrasse 11, 6330 Cham, Switzerland

The book is dedicated to mathematical models of mob control with threshold (conformity) collective decision-making of the agents.

Based on the analysis results of the interconnection between the micro- and macromodels of active network structures, the static (deterministic, stochastic and game-theoretic) and dynamic (discrete and continuous-time) models of mob control are considered. Much attention is given to models of informational confrontation. Many results are applicable not only to mob control problems but also to control problems arising in social groups, online social networks, etc. The book is intended to the researchers and practitioners, as well as the undergraduate and postgraduate students, doctoral candidates specializing in the field of collective behavior modeling.

Contents

Chapter 1
Introduction

Mob is understood below as an active or aggressive gathering of people, i.e. an active or aggressive group, crowd, etc. In scientific literature, *mob control* has several stable and widespread interpretations.

Control of agents' motion (goal achievement, collision avoidance, obstacle avoidance, formation control, etc.). This direction of group control demonstrates intensive development since the early 2000s, embracing two broad fields of research, namely, analytical and simulation (agent-based) models. For instance, *agent-based models* of mob dynamics were surveyed in [2, 14, 86]; a classical example is models of evacuation from buildings. Thousands of papers and tens of reviews were published in these fields. A separate aspect concerns choosing a set of physical influence measures for a mob (prevention of jams, mass riots, and so on), which also represents a subject of many investigations.

Control of mob behavior (*decision-making*). Here it is possible to distinguish between two large directions of research, viz., humanitarian descriptive investigations (within the framework of social psychology, to be more precise, its branch known as *mob psychology* [5, 30, 39, 57, 60, 61, 63, 84, 87]) and mathematical modeling (e.g., see a short survey in [32]). By-turn, the latter is decomposed into two main subdirections as follows.

The first one covers models of *teams* (joint adaptive decision-making by groups of people using information about uncertain factors). In this context, we refer to an overview in [7, 71].

The second subdirection (actually, contributed by the present book) originates from the classical paper [44] and monographs [82, 83] that induced an avalanche of research efforts in the field of mathematical modeling of the so-called *conformity threshold collective behavior*; by assumption, the decision of a given agent depends on the proportion or number of other agents making a corresponding decision (particularly, mob behavior), see the surveys [20, 21, 56, 86]. Nevertheless, despite numerous works dedicated to mob behavior description, there exist few formal statements of mob control problems to date.

© Springer International Publishing AG 2017
V.V. Breer et al., *Mob Control: Models of Threshold Collective Behavior*,
Studies in Systems, Decision and Control 85,
DOI 10.1007/978-3-319-51865-7_1

Mobs are classified using several bases. This book proceeds from a substantial hypothesis that a mob can be united by a common object of attention (be *organized* in some sense), while its members–people (*agents*)—may undertake certain actions or not. To put it tentatively, agents can be *active* (e.g., supporting some decision, participating in mass riots, etc.) or *passive*. Accordingly, mob control represents a purposeful impact (as a rule, informational influence) exerted on a whole mob or separate agents for implementing their desired behavior [74]. If the *goal of control* is to optimize the number or proportion of active agents, exactly this index forms a major control efficiency criterion.

As actions of agents and communications among them are both vital for mob analysis, a mob can be treated as a special case of an *active network structure* (ANS). Other special cases of ANSs are *social groups*, online *social networks* and so on. Some models developed in the book turn out applicable to wider classes of ANSs, not only to mob control problems. In such cases, the term ANS will be used.

The works [69, 70] identified several levels of the description and analysis of ANSs. At level 1 (the lowest one), a network of agents is considered "in toto" using statistical methods, semantic analysis techniques, etc., which makes up *the macromodel of ANS*. At level 2, the structural properties of a network are studied within the framework of graph theory. At level 3, the informational interaction of agents is analyzed, making up *the micromodel of ANS*. Here a researcher disposes of a wide range of applicable models, namely, Markov models (including con- sensus models), finite-state automata, the models of innovations diffusion, infection models, and others; a good survey can be found in the book [48]. At level 4, associated control problems are posed and solved by optimal control or discrete optimization methods, also in terms of the micromodels that reflect the interaction of separate agents. And finally, at level 5, game theory (including reflexive games) is used to describe *informational confrontation*, i.e., the interaction of active sub- jects affecting a social network for their individual goals.

Therefore, each level of the ANS description operates a large set of feasible models and methods—a user toolkit for solving current problems. On the one hand, it is possible to adapt well-known models and methods. On the other hand, the ANS specifics (particularly, the features of a mob as a controlled object) arising at each level require new methods with proper consideration of the high-dimensional dis- tributed incompletely observable object (mob), many interacting subjects pursuing different interests, and so on.

The descriptive heterogeneity of ANSs in terms of various researcher-relevant aspects [69] directly follows from these specifics. At the same time, it would be desirable to surmount difficulties connected with big data [40, 67]. Among them, we mention (a) abstraction as transition from detailed micromodels to macroones involving aggregated characteristics (without loss of substantial information) and (b) formulation of ANS analysis/control problems in terms of macromodels.

The book has the following structure. Chapter 2 summarizes the models of threshold collective behavior underlying the approaches proposed below. Chapter 3 analyzes the interconnection between the micro- and macromodels of ANSs, as well as the issues of their identification. Chapters 4, 5, 6 present static deterministic, stochastic and game-theoretic models of mob control, respectively. Chapters 7 and 8 cover dynamic models of mob control in the discrete- and continuous-time settings, respectively. Chapters 9 and 10 develop the micro- and macromodels of informational confrontation in ANSs, respectively. And finally, Chap. 11 considers models of "spontaneous" mob excitation. The conclusion outlines some promising directions of current and future investigations.

The authors thank A. Yu. Mazurov, Cand. Sci. (Phys.-Math.), for the careful translation of the book from Russian into English, as well as for several helpful remarks improving the presentation.

This work was supported in part by the Russian Scientific Foundation, project no. 16-19-10609.

Chapter 2
Models of Threshold Collective Behavior

Consider the following model of an active network structure that includes several interacting *agents*. Each agent chooses between two decisions resulting in one of two admissible *states*, namely, "1" (*active*, the excited state) or "0" (*passive*, the normal or unexcited state). For instance, possible examples are a *social network* [48] or a *mob* [23], where the active state means participation in mass riots.

While making his decision, each agent demonstrates *conformity behavior*, taking into account the so-called *social pressure* [30, 87] as the observed or predicted behavior of the environment: if a definite number (or proportion) of his "neighbors" are active, then this agent chooses activity. The minimum number (or proportion) of neighbors that "excites" a given agent is called his *threshold*. Note that there also exist models of *anti-conformity* behavior and "mixture" of conformity and anti-conformity, see [26].

Numerous models of *threshold collective behavior* [20, 21] extending Granovetter's basic model [44] define the "equilibrium" state of a mob within collective behavior dynamics via the *distribution function* of agents' *thresholds*. The framework of the game-theoretic models of threshold behavior [18, 19] also treats thresholds' distribution as a major characteristic determining the set of Nash equilibria in the agents' game.

The model adopted below is close to the agent-oriented models such as the bounded-neighborhood model and the spatial proximity model proposed by T. Schelling [82].

If the relationship between the equilibrium state of a system (a social network, a mob) and the threshold distribution function is known, one can pose threshold control problems, e.g., find an appropriate control action modifying agents' thresholds so that the system reaches a desired equilibrium.

Consider the following *model* of a *mob* as a set $N = \{1, 2, \ldots, n\}$ of agents. Agent $i \in N$ is characterized by

(1) the *influence* $t_{ji} \geq 0$ on agent j (a certain "weight" of his opinion for agent j); for each agent j, we have the normalization conditions $\sum_{i \neq j} t_{ji} = 1, t_{ii} = 0$;

© Springer International Publishing AG 2017
V.V. Breer et al., *Mob Control: Models of Threshold Collective Behavior*,
Studies in Systems, Decision and Control 85,
DOI 10.1007/978-3-319-51865-7_2

(2) the decision $x_i \in \{0; 1\}$;

(3) the *threshold* $\theta_i \in [0; 1]$, defining whether agent i acts under a certain *oppo-nents' action profile* (the vector $\bar{x}_{-i} = (x_1, \ldots, x_{i-1}, x_{i+1}, \ldots, x_n)$ comprising the decisions of the other agents except agent i). Formally, define the action x_i of agent i as the best response to the existing opponents' action profile:

$$x_i = BR_i(x_{-i}) = \begin{cases} 1, & \text{if } \sum_{j \neq i} t_{ij} x_j \geq \theta_i \\ 0, & \text{if } \sum_{j \neq i} t_{ij} x_j < \theta_i. \end{cases} \tag{2.1}$$

This book has independent numbering of formulas for each section; while referring to a formula from another section, the double numbering system is used, where the first number indicates the section.

The behavior described by (2.1) is called *threshold behavior*, see surveys in [20, 21]. A *Nash equilibrium* is an agents' action vector $\bar{x}_N$ such that $\bar{x}_N = BR(\bar{x}_N)$, where

$$BR(\bar{x}_N) = (BR_1(\bar{x}_{-1}), \ldots, BR_n(\bar{x}_{-n})).$$

Consider the *following discrete-time dynamic model of collective behavior* [23]. At the initial (zero) step, all agents are passive. At each subsequent step, the agents act simultaneously and independently according to the best-response procedure (2.1).

Introduce the notation

$$Q_0 = \emptyset, Q_1 = \{i \in N \mid \theta_i = 0\},$$

$$Q_k = Q_{k-1} \cup \left\{ i \in N \,\middle|\, \sum_{j \in Q_{k-1}, j \neq i} t_{ij} \geq \theta_i \right\}, \quad k = 1, 2, \ldots, n-1. \tag{2.2}$$

Clearly, $Q_0 \subseteq Q_1 \subseteq \ldots \subseteq Q_n \subseteq N$. Let $T = \{t_{ij}\}$ be the influence matrix of th agents and $\theta = (\theta_1, \theta_2, \ldots, \theta_n)$ correspond to the vector of their thresholds. Evaluate the following index:

$$q(T, \theta) = \min\{k = \overline{0, n-1} \mid Q_{k+1} = Q_k\}. \tag{2.3}$$

Define the *collective behavior equilibrium* (CBE) [23]

$$x_i^*(T, \theta) = \begin{cases} 1, & \text{if } i \in Q_{q(T,\theta)} \\ 0, & \text{if } i \in N \backslash Q_{q(T,\theta)}, i \in N. \end{cases} \tag{2.4}$$

The value

$$x^* = \frac{\#Q_{q(T,\theta)}}{n} = \frac{1}{n}\sum_{i \in N} x_i^*(T,\theta) \tag{2.5}$$

with # denoting set power characterizes the proportion of active agents in the CBE.

Further exposition mostly deals with the *anonymous case* where the graph of agents' relations is complete: $t_{ij} = 1/(n-1)$. In the anonymous case, expression (2.1) takes the form

$$x_i = BR_i(\bar{x}_{-i}) = \begin{cases} 1, & \text{if } \frac{1}{n-1}\sum_{j \neq i} x_j \geq \theta_i, \\ 0, & \text{if } \frac{1}{n-1}\sum_{j \neq i} x_j < \theta_i. \end{cases} \tag{2.6}$$

Designate by $F(\cdot)$: $[0, 1] \rightarrow [0, 1]$ the *distribution function of agents' thresholds*, a nondecreasing function defined on the unit segment that is left continuous and possesses right limit at each point of its domain. Let $\{x_t \in [0, 1]\}_{t \geq 0}$ be a discrete sequence of the *proportions of active agents*, where t indicates time step.

Assume that the proportion x_k of active agents at step k is known ($k = 0, 1, \ldots$). Then the following recurrent expression describes the dynamics of the proportion of active agents at the subsequent steps [19–27, 44, 56]:

$$x_{l+1} = F(x_l), \quad l = k, k+1, \ldots. \tag{2.7}$$

(as a matter of fact, in theory of conformity collective behavior, this equation is sometimes called *Granovetter's behavior*).

The equilibria of system (2.7) are defined by the initial point x_0 (as a rule, $x_0 = 0$) and by the intersection points of the distribution function $F(\cdot)$ with the bisecting line of quadrant I, see [19, 23, 44]:

$$F(x) = x. \tag{2.8}$$

Note that 1 forms a trivial equilibrium due to the properties of the distribution function.

Potentially stable equilibria are points where the curve $F(\cdot)$ crosses the bisecting line approaching it "from left and top."

Denote by $y = \inf\{x : x \in (0, 1], F(x) = x\}$ the least nonzero root of Eq. (2.8). The collective behavior equilibrium (CBE) and, as shown in [23], a Nash equilibrium of the agents' game is the point

$$x^* = \begin{cases} y, & \text{if } \forall z \in [0, y] : F(z) \geq z, \\ 0, & \text{otherwise.} \end{cases} \tag{2.9}$$

According to the properties of the distribution function, for implementing a nonzero CBE a sufficient condition is $F(0) > 0$.

Therefore, given an initial state (the proportion of active agents at step 0), further dynamics of system (2.7) and its equilibrium states depend on the properties of the distribution function of agents' thresholds. Hence, a goal-oriented modification of this function can be treated as *mob control*.

Possible ways of such control that vary the equilibrium states by affecting the parameters of the threshold distribution function will be analyzed in the forthcoming sections.

Chapter 3
Micro- and Macromodels

This section considers two approaches to the design and analysis of ANSs, namely, macro- and microdescriptions [13, 24]. According to the former approach, the structure of relations in a network is averaged, and agents' behavior is studied "in the mean." The latter approach takes into account the structural features of the influence graph of agents and their individual decision-making principles. The first and second approaches are compared using the threshold model of collective behavior with a common relative threshold. And finally, the results of identification and simulation experiments are provided.

3.1 Micromodel

Let $N = \{1, 2, \ldots, n\}$ be a set of *agents* entering an ANS described by a directed graph $R = (N, E)$, where $E \subseteq N \times N$ indicates the arc set. Agents in the network *influence* each other, i.e., arc (i, j) from node i to node j means that agent i trusts agent j. For agent i, denote by $N^{in}(i) = \{j \in N \mid \exists (j; i) \in E\}$ the set of his "*neighbors*" (agents influencing agent i) and by $N^{out}(i) = \{j \in N \mid \exists (i; j) \in E\}$ the set of agents influenced by agent i; $n^{out}(i) = \#N^{out}(i)$ and $n^{in}(i) = \#N^{in}(i)$, where # stands for the power of a set.

The mutual influence of agents can be defined via the direct influence (confidence) matrix $A = \|a_{ij}\|$ having dimensions $n \times n$, where $a_{ij} \geq 0$ characterizes the *confidence* of agent i in agent j (equivalently, the influence of agent j on agent i; $\forall j \notin N^{in}(i) a_{ij} = 0$) [45]. By assumption, we have the *normalization condition*:

$$\forall i \in N: \sum_{j=1}^{n} a_{ij} = 1.$$

If agent i trusts agent j and agent j trusts agent k, then agent k indirectly influences agent i and so on. In other words, there may exist different indirect influence "chains" [47, 48].

© Springer International Publishing AG 2017
V.V. Breer et al., *Mob Control: Models of Threshold Collective Behavior*,
Studies in Systems, Decision and Control 85,
DOI 10.1007/978-3-319-51865-7_3

Suppose that at initial step each agent has an *opinion* on some issue. The opinions of all agents in the network are reflected by a column vector θ^0 of length n, which comprises real-valued nonnegative initial opinions. Possible examples of such opinions are the readiness to support a certain candidate in an election, to purchase a certain product, etc. We refer to the book [48] for a classification and numerous examples of opinions. Agents in the ANS interact via exchanging their opinions. During this process, the opinion of each agent varies under the influence of other agents he trusts. Assume that the opinion $\theta_i^k \in \Re^1$ of agent i at step k [8, 33, 48, 85] has the form

$$\theta_i^k = \sum_{j \in N} a_{ij} \theta_j^{k-1}, \quad k = 1, 2, \ldots. \tag{3.1}$$

Designate by $\theta^k = (\theta_1^k, \theta_2^k, \ldots, \theta_n^k)$ the ANS state at step k.

Suppose that *a consensus is reachable*: as the result of multiple exchange of opinions, the opinions of all agents converge to a common *final* opinion $\theta = \lim_{k \to \infty} \theta^k$. The general necessary and sufficient conditions of such convergence can be found in [31, 78, 85]. In this case,

$$\theta = A^\infty \theta^0, \tag{3.2}$$

where $A^\infty = \lim_{k \to \infty} A^k$. As is well-known (see references in [48]), consensus reachability leads to identical rows in the matrix A^∞. Subsequently, the vector θ consists of identical elements and we will treat it as a scalar. Denote by a_i^∞, $i \in N$, element i of an arbitrary row in this matrix.

For defining the relationship between the common final opinion of all agents and their initial opinions, a possible "alternative" to model (3.2) involves agents' *reputations* [48] $\{r_i \in [0, 1]\}_{i \in N}$, $\sum_{j=1}^n r_j = 1$, i.e., $\theta = \sum_{j=1}^n r_j \theta_j^0$.

Within the framework of micromodel (3.1), it is possible to formulate and solve control problems by affecting initial state, communications among agents, etc. [8, 48].

Now, pass from the micromodel of an ANS with pairwise interaction of the agents to its aggregated description in terms of probabilistic distributions (opinions, reputations, etc.).

3.2 Macromodel

For a given graph G, one can construct empirical probability distributions for the numbers of incoming and outgoing arcs. Designate them by $P^{in}(k)$ and $P^{out}(k)$, respectively, where $k = \overline{0, n-1}$.

Under a given vector of initial agents' opinions, it is possible to find the empirical distribution function of these opinions:

$$F_{\theta^0}(x) = \frac{1}{n} \#\{i \in N \mid \theta_i^0 \leq x\}.$$

Let $P_{\theta^0}(x)$ stand for the corresponding probability distribution.

The macromodel of an ANS is the set $\{n, P^{in}(k), P^{out}(k), P_{\theta^0}(x)\}$; by assumption, the ANS structure guarantees consensus reachability.

There exist several techniques for passing from the microdescription to the macrocharacteristics of ANSs and conversely. In particular, the influence and reputation of agents can be introduced in different ways. For instance, today researchers address two basic (most widespread) models of influence and propagation ("diffusion") of activity (information, opinions, etc.) in ANSs, namely, the linear threshold model (LTM) [44] and the independent cascade model (ICM) [41, 53]. The frameworks of these models consider two major problems in social networks: (3.1) resulting influence maximization (under a limited budget, choose an initial set of excited agents to maximize the resulting excitation) and (3.2) early detection of external impacts (under a limited budget, choose a location of "detectors" in a social network to minimize the resulting influence of external impacts) [29, 45, 53, 73]. For example, the paper [53] demonstrated the following fact based on the analysis results of submodular set functions [66]: the choice problems of the sets of initially excited agents are *NP*-complex in both models. Furthermore, the authors [53] proposed a greedy heuristic $(1-1/e)$-optimal algorithm of their solution.

It is possible to utilize one or more *assumptions* below. The first class of assumptions (R.1–R.3) allows determining the reputations of agents by a given graph G (sometimes, the "influence" of agents is used).

R.1. Agent's reputation in an ANS is directly proportional to the number of agents under his influence, i.e.,

$$r_i = \frac{n^{out}(i)}{\sum\limits_{j \in N} n^{out}(j)}, i \in N. \tag{3.3}$$

R.2. Agent's reputation in an ANS with a reachable consensus is defined by the weight of his initial opinion in the common final opinion:

$$r_i = a_i^\infty, i \in N. \tag{3.4}$$

R.3. Agent's reputation in an ANS is calculated by the PageRank algorithm (e.g., see [58]), i.e., the reputation vector satisfies the system of equations

$$r_i = \sum\limits_{j \in N^{in}(i)} \frac{r_j}{n^{out}(j)}, i \in N. \tag{3.5}$$

Note that the lists of such assumptions are open: other assumptions can be introduced [45] depending on available real data and practical interpretations.

In the mathematical sense, the above class of assumptions (R.1–R.3) admits the following explanation. Reputation (3.3) appears directly proportional to the node degree of the graph G, whereas the reputation vector represents the empirical probability distribution of node degrees. And the reputation vector (3.4) is an invariant distribution of agents' influences; according to expression (3.5), reputation becomes directly proportional to the node degree of the graph G taking into account the "indirect" influences.

The second class of assumptions (I.1, I.2) characterizes the independence of the ANS microparameters in the macrostatistical sense.

I.1. Agent's reputation is independent from his opinion and vice versa.

I.2. The initial opinions of agents are independent, and the initial opinion of agent i does not depend on $N^{in}(i)$ and $N^{out}(i)$.

The third class of assumptions (A.1–A.3) makes it possible to find the direct influence/confidence matrix A under a given graph G and/or given reputations of agents.

A.1. Agent i equally trusts all agents from the set $N^{in}(i)$, i.e., $a_{ij} = \frac{1}{n^{in}(i)}$, $i \in N$, $j \in N^{in}(i)$.

A.2. The confidence of agent i in agent $j \in N^{in}(i)$ is directly proportional to the latter's reputation, i.e., $a_{ij} = \sum_{k \in N^{in}(i)} \frac{r_j}{r_k}$, $i \in N$, $j \in N^{in}(i)$.

Therefore, by making certain assumptions on the general properties of ANSs, one can establish quantitative relations between their micro- and macromodels.

Finalizing this brief discussion of the ANS macromodels, we emphasize an important aspect. An adequate mathematical apparatus of their analysis consists in *random graph* theory (e.g., see [15, 35, 65]) pioneered by P. Erdos and A. Renyi [38]. Actually, this framework is efficiently applied not just to social networks, but also to telecommunication, informational, technological, biological, genic, artificial neural, and other networks, scientific community networks, etc. (numerous examples can be found in [3, 35, 65]).

Imagine that an arc between any pair of nodes in an ANS graph exists or does not with an identical probability for any pair (this property and similar properties can be derived from modeling of graph formation dynamics, see [3, 36, 37]). Then we have the binomial distribution of the number of incident arcs (in the limit case— under a large number of graph nodes—the Poisson distribution). Such graphs are called *exponential* or *Erdos–Renyi graphs*.

Interestingly, the numbers of incident arcs in most social networks (and in the World-Wide Web) obey not exponential, but heavy-tailed *power-series "distributions"* $P^{in}(k)$ and $P^{out}(k)$ [3, 11, 12]. Below we employ exactly these distributions ($P^{in}(k) \sim k^{-\gamma}$, where $1 < \gamma < 4$). Given a graph defined by the statistical characteristics of its basic parameters, one can find the macrocharacteristics (probability distributions) of influence, reputation, confidence, etc. based on the above-stated assumptions.

3.3   Threshold Model of Agents Behavior

Let us examine the equivalence of micro- and macrodescriptions from the theoretical viewpoint using the threshold model of collective behavior of agents in ANSs.

The models of agents' opinions dynamics in ANSs (see above) involve a single characteristic of each agent—his opinion—and the rest parameters reflect the interaction of agents. The so-called *behavioral models of ANSs* are richer: in addition to "internal" parameters, they incorporate variables describing agent's behavior (his *decisions*). Generally, these decisions depend on the internal parameters of an agent (his opinions, individual characteristics) and, may be, on the opinions and/or actions of other agents (all agents, neighbors or some group of agents). As an example of behavioral models, we choose the threshold model (the general game-theoretic modeling approaches to collective threshold behavior were outlined in the publications [18, 19]).

Consider a social network composed of a set $N = \{1, 2, \ldots, n\}$ of *agents*. Each of them chooses between two options, namely, *being active* or *being passive*. Denote by $x_i \in \{0; 1\}$ the choice of agent i, where $x_i = 1$ means that the agent is active (if $x_i = 0$, passive). The choice of agent i is influenced by a set $D_i = N^{in}(i)$ of other agents—his neighbors. Notably, agent i decides to be active or passive depending on his threshold $\theta_i \in [0, 1]$ and the proportion of active neighbors: if more than $\theta_i \mid D_i \mid$ neighbors are active, agent i follows their choice.

We believe that the opinion of a separate agent is his individual threshold $\theta_i \in [0, 1]$. Moreover, the ANS structure admits a consensus, i.e., there exists a common opinion θ characterizing all agents of the network in the final analysis (see formula (3.3)). In the sequel, this common opinion will be called *the common relative threshold* of the agents.

The microlevel behavior of an agent can be expressed by the best response (also, see expression (2.1))

$$x_i = BR_i(x_{-i}) = \begin{cases} 1, & \sum_{j \in D_i} x_j > \theta \, d_i, \\ 0, & \sum_{j \in D_i} x_j \le \theta \, d_i, i \in N, \end{cases} \qquad (3.6)$$

Here $d_i = |D_i|$ gives the number of neighbors of agent $i \in N$. The behavioral model defined by (3.6) is *the micromodel with the common relative threshold*.

Now, pass to the probabilistic macrodescription of the threshold behavioral model. Assume that agents are undistinguishable and the number of agent's neighbors makes a random positive integer $d : 1 \le d \le n - 1$. Our analysis ignores the networks of agents without neighbors ($d = 0$): such agents are always passive due to (3.6). Let $M(d) = P^{in}(d) : \{1, 2, \ldots, n - 1\} \to [0, 1]$ be the probability that the number of neighbors is d. Consider the averaged dynamics of agents' interaction in discrete time. Suppose that q agents are active at an arbitrary step, and find the expected number of active agents at subsequent step within micromodel (3.6).

First, calculate the probabilities G_n of the events that exactly k *agent's neighbors are active*. An agent has d neighbors in C_{n-1}^d possible combinations. The number of combinations where exactly $k \leq d$ neighbors of an agent are active makes up C_q^k, since the network contains q active agents totally. Similarly, C_{n-1-q}^{d-k} gives the number of combinations where exactly $d - k$ neighbors of an agent are passive ($n - 1 - q$ is the total number of passive agents in the network). According to combinatorial calculus, the probability that exactly k of d agent's neighbors are active obeys the hypergeometric distribution:

$$G_n(q, d, k) = \frac{C_q^k C_{n-1-q}^{d-k}}{C_{n-1}^d}. \tag{3.7}$$

Evaluate the probability P_n that an agent becomes active under the influence of q active agents in the network. For an agent to be active, a necessary condition is that more than θd his neighbors are active, see (3.6). The desired probability $P_n = P_n(q, d, \theta)$ represents the sum of the probabilities (3.7) over all k: $[\theta d] < k \leq d$ (as before, $[\cdot]$ denotes the integer part operator):

$$P_n(q, d, \theta) = \sum_{k=[\theta d]+1}^{d} G_n(q, d_i, k) = 1 - \sum_{k=0}^{[\theta d]} \frac{C_q^k C_{n-1-q}^{d-k}}{C_{n-1}^d}. \tag{3.8}$$

Probability (3.8) equals the proportion of active agents at subsequent step among the ones having d neighbors. Hence, the expected proportion of active agents at subsequent step is

$$F_n(q, \theta) = \sum_{d=1}^{n-1} P_n(q, d, \theta) M(d). \tag{3.9}$$

And the number of active agents in the network evolves according to the recurrent scheme

$$q_{k+1} = [n F_n(q_k, \theta)]. \tag{3.10}$$

To proceed, we explore the behavior of the function (3.9) for sufficiently large n. In this case, the hypergeometric distribution (3.7) is approximated by the binomial distribution with the probability $p = q/n$:

$$G_n(q, d, k)|_{n \gg 1, p=q/n} \approx b(p, d, k) = C_d^k p^k (1 - p)^{d-k}.$$

By analogy with (3.8), the probability that more than θd agent's neighbors are active is

$$P_n(q,d,\theta)|_{n \gg 1, p=q/n} \approx B(p,d,\theta) = 1 - \sum_{k=0}^{[\theta d]} b(p,d,k).$$

Then the distribution of the number of active agents (3.9) and the dynamics of the proportion of active agents in the ANS can be rewritten as

$$F_n^B(p,\theta) = \sum_{d=1}^{n-1} B(p,d,\theta)M(d), \tag{3.11}$$

$$p_{k+1} = F_n^B(p_k,\theta). \tag{3.12}$$

The behavioral model (3.12) is the *macromodel with the common relative threshold*.

Threshold models research was initiated by the classical work [44]; we outline the key results derived by M. Granovetter. All agents are the neighbors of each other (the relation graph R appears complete), and the number of agents is not specified. Each agent possesses a threshold—if the proportion of active agents exceeds this quantity, the agent becomes active. Moreover, the value of the threshold is described by a distribution function F. Assume that the proportion of active agents at step k equals r_k. Hence, all agents whose thresholds are smaller than r_k ($F(r_k)$ agents totally) choose being active at subsequent step. Therefore, we obtain the recurrent formula

$$r_{k+1} = F(r_k). \tag{3.13}$$

Macromodel (3.12) with the common relative threshold is equivalent to Granovetter's model (3.13) in the following sense. Suppose that we know the distribution (3.11). Then it is possible to construct the corresponding Granovetter's model by setting $F(p) = F_n^B(p,\theta), p \in [0,1]$. Conversely, imagine that Granovetter's model with the threshold distribution function F is given. Solve numerically Eq. (3.11) for $M(\cdot)$ (solution of such equations forms an independent problem not considered here). The distribution M completely characterizes the macromodel with the common relative threshold.

The paper [19] demonstrated that the equilibrium conditions in Granovetter's model ($F(x) = x$) are equivalent to the Nash equilibrium conditions in the micromodel with the common relative threshold provided that the influence graph of agents enjoys completeness: $d_i = n - 1, \forall i \in N$. In other words, the macromodel of agents' threshold behavior in an ANS being available, one can easily find the equilibrium states in this network.

This subsection has established the theoretical connection between the micro- and macromodels of ANSs. Identification of the suggested approaches and mutual adequacy of the micro- and macrodescriptions are studied "experimentally" below and in the paper [13] for some online social networks. A promising direction of further research concerns constructing and analyzing the thermodynamics and statistical physics interpretations of the macromodels of ANSs (see the papers [1, 79] and the surveys [20, 21, 86]). Another line of future investigations includes statement and solution of associated control problems (e.g., by analogy with the control problems of ANSs [8–10, 48, 73] described by expressions (3.1) or (3.6)).

3.4   Identification and Simulation Experiments

This subsection, written jointly with A.V. Batov, focuses on identification of the micro- and macrocharacteristics of ANSs proposed above, employing data on real online social networks—*Facebook*, *LiveJournal* and *Twitter*. The results of corresponding simulation experiments are provided and compared [13].

In the previous subsections, we have introduced the notions of the *micro-* and *macromodels of a social network* with a *common relative threshold* $\theta \in [0, 1]$.

Within the framework of the *micromodel*, the game-theoretic approach describes agent's behavior via his best response BR_i. At each step k, agents make the following decisions simultaneously and independently from each other:

$$x_i^{(k)} = BR_i\left(x_{-i}^{(k-1)}\right) = \begin{cases} 1, & \sum_{j \in D_i} x_j^{(k-1)} > \theta \, d_i, \\ 0, & \sum_{j \in D_i} x_j^{(k-1)} \leq \theta \, d_i. \end{cases}, i \in N = \{1, 2, \ldots, n\}. \quad (3.14)$$

For agent i, the quantity $d_i = |D_i|$ denotes the number of his neighbors, D_i is the set of his neighbors and $x_{-i}^{(k-1)} = \left\{x_1^{(k-1)}, x_2^{(k-1)}, \ldots, x_{i-1}^{(k-1)}, x_{i+1}^{(k-1)}, \ldots, x_n^{(k-1)}\right\}$ forms the opponents' action profile. Here $x_i^{(k)} \in \{0; 1\}$, and $x_i^{(k)} = 1$ ($x_i^{(k)} = 0$) means that the agent is active (passive, respectively) at step k.

The macromodel describes the dynamics of the proportion $p_k = \frac{1}{n}\sum_i x_i^{(k)} \in [0, 1]$ of active agents:

$$p_{k+1} = F_n(p_k, \theta), \quad (3.15)$$

where

$$F_n(p, \theta) = \sum_{d=1}^{n-1} B(p, d, \theta)M(d), \quad (3.16)$$

$B(p, d, \theta) = 1 - \sum_{k=0}^{[\theta d]} C_d^k p^k (1-p)^{d-k}$ is the binomial distribution function, and $M(d)$ indicates the probabilistic distribution of the number of neighbors d in an ANS graph (for details, see above).

Verification and comparison of the theoretical models (3.14) and (3.15) proceed from data on real online social networks (SNs). In this work, we employ simulation methods for micromodel (3.14) and different-type approximations of the distribution function (3.16) in macromodel (3.15).

The relations among agents in a real SN can be reflected by a directed graph G. The direction of an edge from one agent (node) to another shows the former's influence on the latter. The micromodels address explicitly the influence graph, whereas the macromodels operate its macrocharacteristic, i.e., the distribution $M(\cdot)$ of the number of neighbors. The applicability conditions of macromodel (3.15) dictate that the number of agents is sufficiently large. Therefore, we have analyzed influence relations in three large SNs, viz., the Russian-language segments of *Facebook* (*F*), *LiveJournal* (*L*) and *Twitter* (*T*).

For instance, in *Facebook* an agent has connections to his friends, which can be interpreted as the influence relations of these friends on the agent. In *LiveJournal* and *Twitter*, directed influence relations are agent's subscriptions for viewing and commenting information posted by other agents. We will believe that all agents influencing a given agent in a network are his *neighbors*, see expression (3.1).

Table 3.1 contains the macroindicators of these SNs: the maximum number of neighbors (*MaxFrnds*), the number of agents (*Users*), the number of agents with nonzero amounts of neighbors (*Nonzero users*), the total number of relations (*Links*) and the average number of neighbors for agents with nonzero amounts of neighbors (*AvgF = Links/Nonzero users*).

According to Table 3.1, the number of agents in the SNs is large, which enables hypothesizing about the applicability of macromodel (3.15).

The distribution function (3.16) of macromodel (3.15) includes two components, namely, (a) the probability $B(p, d, \theta)$ that a given proportion p of d agents is active and (b) the distribution $M(\cdot)$ of the number of neighbors in an SN. Actually, these functions can be defined in different ways, which generates the following group of problems.

Problem 3.1 Identification of the distribution functions $M(\cdot)$ in the three SNs. Here we construct the empirical distributions $M_F(\cdot)$, $M_L(\cdot)$ and $M_T(\cdot)$, as well as find analytically their approximating functions $\overline{M}_F(\cdot)$, $\overline{M}_L(\cdot)$ and $\overline{M}_T(\cdot)$.

Problem 3.2 Design and analysis of the simulation models of threshold behavior specified by the best response (3.14). By assumption, randomly chosen agents are active at the initial step. Formula (3.14) serves for evaluating the number of active agents at the subsequent step. Next, we average the result over the random sets of

Table 3.1 The macroindicators of the SNs

SN	MaxFrnds	Users	Nonzero users	Links	AvgF
Facebook	4 199	3 250 580	3 084 017	77 639 757	50.35
LiveJournal	2 499	5 758 706	3 586 959	124 729 288	34.77
Twitter	759 313	~41 700 000	35 427 738	1 418 363 662	40.04

initially chosen agents. The described procedure yields a family of functions (which depends on the parameter θ) to-be-compared with other distribution functions obtained by solving other problems (see Table 3.5).

Problem 3.3 Approximation of the relationships derived via simulation by sigmoids (see Problem 3.2).

Problem 3.4 Definition of a family of the distribution functions (3.16) dependent on the parameter θ. Then, we substitute the empirical distribution functions $M_F(\cdot)$, $M_L(\cdot)$ and $M_T(\cdot)$ of the node degrees of the relation graph into the members of this family instead of $M(\cdot)$.

Problem 3.5 appears similar to Problem 3.4, except that the empirical distribution functions of the SNs are replaced by their approximations $\overline{M}_F(\cdot)$, $\overline{M}_L(\cdot)$ and $\overline{M}_T(\cdot)$ (see Problem 3.1).

Problem 3.6 Comparison of the solution results in Problems 3.2–3.5.

To solve Problems 3.2–3.5, we address two methods, viz., empirical data analysis and their analytical approximation. The general scheme of the study is explained by Table 3.2.

Problem 3.7 Exploration of the collective behavior equilibria dependence on the common relative threshold θ within Granovetter's model constructed for these SNs.

Let us describe the solution of each posed problem.

Identification of the distribution functions $M(\cdot)$ for the number of agents' neighbors in ANSs (Problem 3.1). Today, there exist numerous studies of online SNs testifying that the distribution of the number of neighbors (the distribution of node degrees in large SNs) is well approximated by the power-series distribution (e.g., see [3, 12, 11]). These functions and their approximations for the above SNs are shown in Fig. 3.1. *The curves of the empirical distributions for the number of agent's neighbors in social networks and their linear approximations:*

(a) Facebook; (b) LiveJournal; (c) Twitter. In the log-log scale, here readers can find the curves of the empirical distributions $M_F(\cdot)$, $M_L(\cdot)$ and $M_T(\cdot)$ of the numbers of neighbors. Since a power function in the log-log scale represents a straight line with slope a and the zero-point value of b, we have constructed the best linear approximation. The obtained values of the approximation coefficients for different SNs are combined in Table 3.3. Other notation is explained below.

Table 3.2 The models and methods used in Problems 3.2–3.5

Model	Method	
	Empirical data analysis	Analytical approximation
Micromodel of SN	Problem 3.2	Problem 3.3
Macromodel of SN	Problem 3.4	Problem 3.5

For a small number of neighbors, the distribution of node degrees in the graph G admits approximation by a horizontal line. This leads to the "cut" linear approximations $\overline{M}_F(\cdot)$, $\overline{M}_L(\cdot)$ and $\overline{M}_T(\cdot)$, see Fig. 3.2 and Table 3.4.

For a small number of neighbors, the horizontal line has been chosen due to the following reasons:

- according to the normalization condition, the area under a distribution curve is unity. Direct normalization varies the coefficients of the power-series distribution, thereby affecting the accuracy of approximation;
- as against other approximations, the "cut" linear approximation leads to smaller R-squared. For comparison, we have performed approximation by the Pareto distribution (with the parameter a_pareto from Table 3.3), which makes a line in the log-log scale. The "cut" approximation yields a better result for the empirical distribution, see Table 3.4.

The value c where the linear horizontal approximation becomes "inclined," i.e.,

$$\overline{M}(d) = \begin{cases} \exp[b] \times c^a, d \leq c, \\ \exp[b] \times d^a, d > c, \end{cases}$$

(see c_real in Table 3.3) has been found from the normalization condition $\sum_{d=1}^{n} \overline{M}(d) = 1$.

We have approximated the distribution of node degrees in the SNs. These results will be used in Problem 3.4. Now, let us simulate agents' behavior in micromodel (3.14).

Design and analysis of the simulation models of threshold behavior defined by the best response (3.14) (Problem 3.2). Simulation runs as follows. Consider SNs *Facebook* and *LiveJournal* described by their relation graphs. "Excite" randomly q agents (a proportion $q/n \in [0; 1]$). Next, employ formula (3.14) for evaluating the best response of each agent (being active or passive). According to (3.15), the resulting proportion of active agents is the value $F_n(q/n, \theta)$. Repeat the experiment very many times for different values q belonging to the segment $[0, 1]$. In all trials, the relative deviation of the value $F_n(q/n, \theta)$ has been about 0.001 (this is due to the randomly chosen set of initially excited agents). Figure 3.3 demonstrates the curves of $F_n(q/n, \theta)$ under different values of the parameter θ.

We have obtained the simulation results for SNs *Facebook*, *LiveJournal* and *Twitter*. The next stage is dedicated to their approximation.

Table 3.3 The approximation coefficients of the functions $M_F(\cdot)$, $M_L(\cdot)$ and $M_T(\cdot)$

Function	a	b	c_real	a_pareto
$M_F(\cdot)$	−2.181	3.274	26.628	0.688
$M_L(\cdot)$	−2.208	2.828	16.878	0.765
$M_T(\cdot)$	−1.802	−0.196	1.8233	0.799

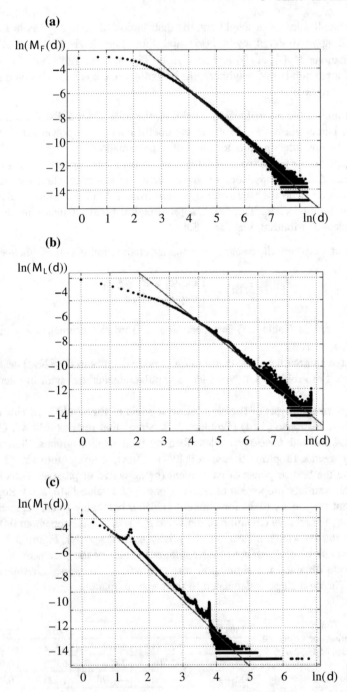

Fig. 3.1 The curves of the empirical distributions for the number of agent's neighbors in social networks and their linear approximations: **a** *Facebook*; **b** *LiveJournal*; **c** *Twitter*

(a)

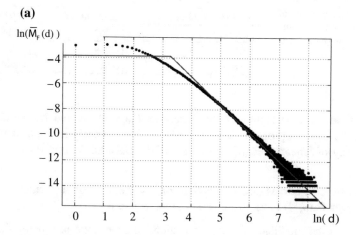

(b)

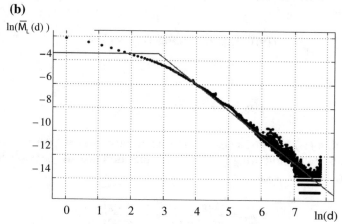

(c)

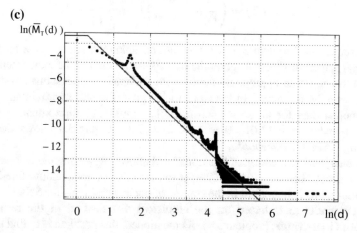

Fig. 3.2 The curves of the "cut" linear approximation of the function M(·): **a** *Facebook*; **b** *LiveJournal*; **c** *Twitter*

Approximation of SN	"Cut" linear	Pareto
Facebook	0.962	0.916
LiveJournal	0.929	0.884
Twitter	0.849	0.849

Table 3.4 The accuracy of the "cut" linear and Pareto approximations (R-squared)

Analytical approximation of the functions $F_n(p, \theta)$ **yielded by simulation** (Problem 3.3). In this problem, it is necessary to find the analytical approximation of the family of functions $F_n(p, \theta)$ for each of the SNs. Direct observation shows that:

- the constructed curves (see Fig. 3.4) belong to the class of sigmoids;
- the curves $F_n(p, \theta)$ have an inflection point at $p \approx \theta$.

And so, as the candidates for approximation we choose the parametric families of functions

$$f(p, \theta, \alpha, \lambda, \gamma) = \alpha \arctg(\lambda(p - \theta)) + \gamma$$

and

$$g(p, \theta, \alpha, \lambda, \gamma) = \frac{\alpha}{1 + e^{-\lambda(p-\theta)}} + \gamma.$$

Recall that $f(\cdot)$ and $g(\cdot)$ must be distribution functions. Having this aspect in mind, we adopt the following parametric families of functions (here λ acts as the parameter):

$$f(p, \theta, \lambda) = \frac{\arctg(\lambda(p - \theta)) + \arctg(\lambda\theta)}{\arctg(\lambda(1 - \theta)) + \arctg(\lambda\theta)},$$

$$g(p, \theta, \lambda) = \left(\frac{1 - e^{-\lambda p}}{1 + e^{-\lambda(q - \theta)}}\right)\left(\frac{1 + e^{-\lambda(1-\theta)}}{1 - e^{-\lambda}}\right). \tag{3.17}$$

Consequently, Problem 3.3 is (a) to find an unknown parameter λ so that the surface $f(p, \theta, \lambda)$ or $g(p, \theta, \lambda)$ gives the best approximation of the experimental data and (b) to choose an appropriate family with the smallest approximation error.

Interestingly, the family of functions (3.17) gives the best approximation of the experimental data for all social networks. The minimum approximation error is achieved under $\lambda_F = 13.01$, $\lambda_L = 9.18$, $\lambda_T = 7.34$. Figure 3.4 shows the corresponding curve for *Facebook*.

The analytical expression of function (3.16) allows suggesting the one-parameter model of agents' behavior (under different values of the common relative threshold θ). Particularly, this is important in the context of control problems for SNs.

Macromodel (3.2) based on the empirical distribution of the number of neighbors in a graph (Problem 3.4). As mentioned, this problem is to find a family

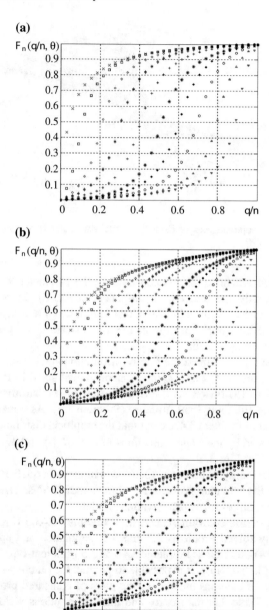

Fig. 3.3 The results of threshold behavior simulation: **a** *Facebook*; **b** *LiveJournal*; **c** *Twitter*. Notation:

× − θ = 0; □ − θ = 0.1; ◇ − θ = 0.2; ╪ − θ = 0.3; + − θ = 0.4;

* − θ = 0.5; ◆ − θ = 0.6; ○ − θ = 0.7; △ − θ = 0.8; ▽ − θ = 0.9

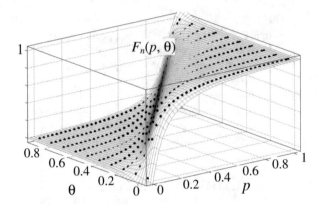

Fig. 3.4 Facebook: Approximation of the experimental data $F_n(p,\theta)$ (*points*) by the analytical family $f(p,\theta,\lambda_F)$ (*grid*)

of the distribution functions (3.16) dependent on the parameter θ; these functions incorporate explicitly the theoretical component $B(p,d,\theta)$. In other words, we substitute the empirical distribution functions $M_F(\cdot), M_L(\cdot)$ and $M_T(\cdot)$ of the node degrees of the relation graph into formula (3.16) instead of $M(\cdot)$. Figure 3.5 presents the corresponding results.

Therefore, we have obtained the family of functions (3.16) for the SNs. Next, study them using the "cut" linear approximations constructed in Problem 3.1.

Macromodel (3.15) based on the distribution of the number of neighbors approximated by the analytical function (Problem 3.5). As a matter of fact, this problem is similar to Problem 3.4, except that the empirical distribution functions of the SNs are replaced by their approximations $\overline{M}_F(\cdot), \overline{M}_L(\cdot)$ and $\overline{M}_T(\cdot)$. The results are demonstrated by Fig. 3.6.

Obviously, the family of functions (3.16) is analogous in qualitative terms to the one obtained in the previous problem for corresponding SNs. The rigorous comparison of these families takes place within Problem 3.6.

Comparison of the solution results in Problems 3.2–3.5 (Problem 3.6). We have successfully solved Problems 3.2 and 3.3 (simulation and approximation of micromodel (3.14)), as well as Problems 3.4 and 3.5 (different-type approximations of macromodel (3.15)). Now, it is possible to parallel the outcomes, see Table 3.5. Here the headers of columns answer for the pairs of compared problems.

According to Table 3.5, the macro- and microdescriptions yield similar results (see Problems 3.2, 3.3 and 3.4). Yet, the closeness of the outcomes for Problems 3.2, 3.3, 3.4 and 3.5, is smaller. A promising line of future research lies in suggesting another function with better approximation of the distribution function $M(\cdot)$ for the node degrees of the relation graph.

Equilibria analysis in the ANSs (Problem 3.7). As established above, agents' behavior in the model with the common relative threshold appears equivalent to threshold behavior in Granovetter's model. Within the latter framework, an

(a)

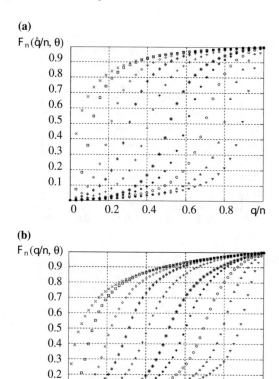

(b)

(c)

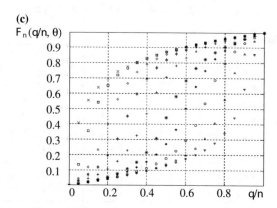

Fig. 3.5 The macromodel (3.15) with the empirical distribution of the number of neighbors in the graph: **a** *Facebook*; **b** *LiveJournal*; **c** *Twitter*;

$\times$ $-\theta = 0$; $\square$ $-\theta = 0.1$; $\diamond$ $-\theta = 0.2$; $\ddagger$ $-\theta = 0.3$; $+ -\theta = 0.4$;
$*$ $-\theta = 0.5$; $\blacklozenge$ $-\theta = 0.6$; $\circ$ $-\theta = 0.7$; $\triangle$ $-\theta = 0.8$; $\triangledown$ $-\theta = 0.9$

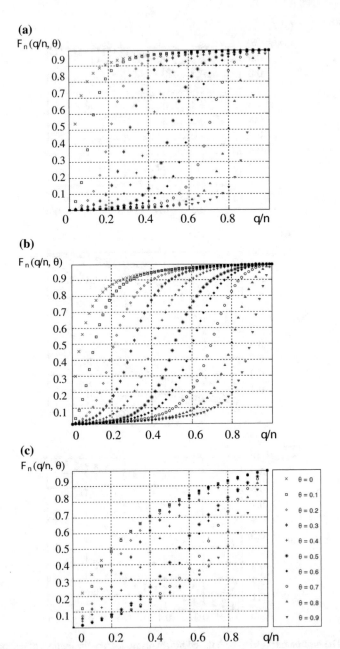

Fig. 3.6 The macromodel (3.15) with the approximated distribution function: **a** *Facebook*;
b *LiveJournal*; **c** *Twitter*;

✗   − θ = 0; ☐ − θ = 0.1; ◇ − θ = 0.2; ✢ − θ = 0.3; ✛ − θ = 0.4;
✳ − θ = 0.5; ◆ − θ = 0.6; ○ − θ = 0.7; △ − θ = 0.8; ▽ − θ = 0.9

Table 3.5 Comparison of the solution results in Problems 3.2–3.5 (R-squared)

SN	Problems 3.2 and 3.4	Problems 3.2 and 3.5	Problems 3.4 and 3.5	Problems 3.2 and 3.3	Problems 3.3 and 3.4	Problems 3.3 and 3.5
Facebook	0.9976	0.9932	0.9911	0.9973	0.9973	0.9907
LiveJournal	0.9999	0.9872	0.9872	0.9960	0.9960	0.9855
Twitter	0.9998	0.9631	0.9642	0.9949	0.9950	0.9599

important role belongs to such properties of equilibria (i.e., points characterized by the equation $F_n(p, \theta) = p$) as their number and stability (crossing the bisecting line of quadrant I in an equilibrium "from left and top"). Let us investigate these issues for the SNs above. By observing the curves in Figs. 3.3, 3.4, 3.5 and 3.6, we naturally arrive at the following conclusion. Depending on the parameter θ, there may exist different sets of equilibria (in the sense of their number, stability or instability, etc.).

The intersection point of the curve $F_n(p, \theta)$ and the diagonal of the unit square lying within the interval $(0, 1)$ is an unstable equilibrium. Really, the curve $F_n(p, \theta)$ crosses the diagonal "from left and top."

Under $\theta \in [\sim 0.1; \sim 0.9]$, this property holds for all social networks considered, see Fig. 3.3. Moreover, the points $q = 0$ and $q = 1$ form stable equilibria.

In the case of $\theta \leq 0.1$, the system possesses two equilibria, viz., the points $q = 0$ (unstable) and $q = 1$ (stable).

Similarly, if $\theta \geq 0.9$, the system admits the same equilibria, but $q = 0$ is stable, whereas $q = 1$ appears unstable.

Figure 3.7 also serves for evaluating equilibria in the recurrent procedure (3.16) of the macromodel under different relationships between the initial state p_0 and the common relative threshold θ. For instance, if the vector (p_0, θ) belongs to domain II in Fig. 3.7, then process (3.16) terminates in the equilibrium $p = 0$. On the other

Fig. 3.7 The relationship between the parameter θ and the intersection point of the curve $F_n(p, \theta)$ with the diagonal of the unit square: —○— • Facebook;—□— • LiveJournal; —◇—• Twitter

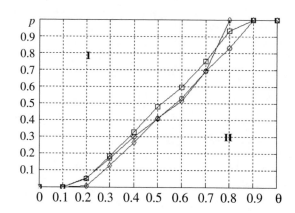

hand, if this vector lies in domain I, process (3.16) ends in the equilibrium $p = 1$ (Fig. 3.7).

The above theoretical macromodels of ANSs have been analyzed through simulation using real data and approximation of the obtained results. The outcomes lead to the following important conclusions.

First, the probabilistic description of the macromodel agrees with the macrodescription: the simulation results well match the results of calculations based on the probabilistic model and the real distribution of the node degrees of relations graphs for different social networks.

Second, despite appreciable differences in the scale and structure of relations graphs for the real social networks, their macromodels (3.15) have much in common: they possess the form of sigmoids and admit good approximation by the parametric family of functions (3.17) with different values of the coefficient λ.

Chapter 4
Deterministic Models of Mob Control

This section studies a threshold behavior model for a group of agents. Making binary decisions (choosing between active or passive states), agents take into account the choice of other members of the group. Control problems for thresholds and agents' reputation are stated and solved in order to minimize the number of agents choosing "to be active" [23].

4.1 A Threshold Model of Mob Behavior

Consider the following model of a mob. There is a set $N = \{1, 2, \ldots, n\}$ of *agents* choosing between two *decisions*, "1" (being active, e.g., participating in mass riots) or "0" (being passive). Agent $i \in N$ is characterized by

- the *influence* on agent j, denoted by $t_{ji} \geq 0$ (a certain "weight" of his opinion for agent j); for each agent j, we have the normalization conditions $\sum_{i \neq j} t_{ji} = 1$, $t_{ii} = 0$;
- the decision $x_i \in \{0; 1\}$;
- the *threshold* $\theta_i \in [0, 1]$, defining whether agent i acts under a certain *opponents' action profile* (the vector x_{-i} comprising the decisions of the rest agents). Formally, define the action x_i of agent i as the best response to the existing opponents' action profile:

$$x_i = BR_i(x_{-i}) = \begin{cases} 1, & \text{if } \sum_{j \neq i} t_{ij} x_j \geq \theta_i, \\ 0, & \text{if } \sum_{j \neq i} t_{ij} x_j < \theta_i. \end{cases} \qquad (4.1)$$

© Springer International Publishing AG 2017
V.V. Breer et al., *Mob Control: Models of Threshold Collective Behavior*,
Studies in Systems, Decision and Control 85,
DOI 10.1007/978-3-319-51865-7_4

The behavior described by (4.1) is called *threshold behavior*. A *Nash equilibrium* is an agents' action vector x_N such that $x_N = BR(x_N)$ [64].

By analogy to the paper [18], adopt the *following dynamic model of collective behavior*. At an initial step, all agents are passive. At each subsequent step, agents act simultaneously and independently according to procedure (4.1). Introduce the notation

$$Q_0 = \{i \in N | \theta_i = 0\},$$

$$Q_k = Q_{k-1} \cup \left\{ i \in N \Big| \sum_{j \in Q_{k-1}, j \neq i} t_{ij} \geq \theta_i \right\}, \quad k = 1, 2, \ldots, n - 1. \tag{4.2}$$

Clearly, $Q_0 \subseteq Q_1 \subseteq \cdots \subseteq Q_{n-1} \subseteq Q_n = N$. Let $T = \{t_{ij}\}$ be the influence matrix of agents and $\theta = (\theta_1, \theta_2, \ldots, \theta_n)$ form the vector of their thresholds. Evaluate the following index:

$$q(T, \theta) = \min\{k = \overline{0, n-1} | Q_{k+1} = Q_k\}. \tag{4.3}$$

Define the collective behavior equilibrium x^* (CBE) by

$$x_i^*(T, \theta) = \begin{cases} 1, & \text{if } i \in Q_{q(T,\theta)}, \\ 0, & \text{if } i \in N \backslash Q_{q(T,\theta)}, \end{cases} \quad i \in N. \tag{4.4}$$

Assertion 4.1 For any influence matrix T and agents' thresholds θ, there exists a unique CBE (4.4) representing a Nash equilibrium in the game with the best response (4.1).

Proof of Assertion 4.1. To establish the existence, one should actually demonstrate the following: the set used for minimization in (4.3) is nonempty. By *argumentum ex contrario*, suppose emptiness of this set. In other words, the sequence of sets $Q_0 \subseteq Q_1 \subseteq \cdots \subseteq Q_{n-1} \subseteq Q_n$ is assumed to have no coinciding elements. This implies that each consequent set differs from the previous one (at least) by a single element. On the other hand, the sequence has $n + 1$ sets, but there are n totally. We have arrived at a contradiction.

Uniqueness follows from the CBE definition—see (4.4)—and from uniqueness of index (4.3).

Let $x^*(T, \theta)$ specify the CBE. And so, all agents belonging to the set $Q_{q(T,\theta)}$ are active. However, according to formulas (4.1)–(4.2), this choice matches their best responses. All agents in the set $N \backslash Q_{q(T,\theta)}$ turn out passive. By virtue of (4.2)–(4.3), these agents satisfy $\sum_{j \in Q_{q(T,\theta)}} t_{ij} < \theta_i$, $i \in N \backslash Q_{q(T,\theta)}$. Then being passive is the best response (see expression (4.1)). Hence, for all i we obtain $x_i = BR_i(x_{-i})$, and $x^*(T, \theta)$ represents a Nash equilibrium. Proof of Assertion 4.1 is complete.• (here and in the sequel, symbol • indicates the end of a proof, example, etc.).

We underline that the above CBE definition possesses constructive character, as its evaluation based on (4.2)–(4.4) seems easy. Moreover, a reader should observe an important fact: without agents having zero thresholds, passivity of all agents makes up the CBE. In the sense of control, this means that most attention should be paid to the so-called "*ringleaders*," i.e., agents deciding "to be active" even when the rest remain passive.

The model with reputation. Denote by $r_j = \frac{1}{n-1} \sum_{i \neq j} t_{ij}$ the average influence of agent $j \in N$ on the rest agents. The quantity r_j is said to be the relative *reputation* of agent j (a certain "weight" of his opinion for the other agents). The other agents consider his opinion or actions according to this weight. Within the framework of the model with reputation, influence can be characterized by the vector $r = \{r_i\}_{i \in N}$.

In this model, define the action x_i of agent i as the best response to the opponents' action profile:

$$x_i = BR_i(x_{-i}) = \begin{cases} 1, & \text{if } \sum_{j \neq i} r_j x_j \geq \theta_i, \\ 0, & \text{if } \sum_{j \neq i} r_j x_j < \theta_i \end{cases}$$

A special case of the described model is the *anonymous case*, where all agents have the identical reputations $r_i = \frac{1}{n-1}$. Then choose integers $m_1, m_2, \ldots, m_n$ as the thresholds and construct the corresponding threshold vector m. Next, sort the agents in the nondecreasing order of their thresholds: $m_1 \leq m_2 \leq \cdots \leq m_n$. Believing that $m_0 = 0$ and $m_{n+1} > n$, define the number $p(m) \in \{0, \ldots, n\}$ by

$$p(m) = \min\{k \in N \cup \{0\} | m_k \leq k, m_{k+1} > k\}.$$

Consequently, the CBE acquires the following structure: $x_i^* = 1$, $i = \overline{1, p(m)}$; $x_i^* = 0$, $i = \overline{p(m)+1, n}$. That is, the first $p(m)$ agents are active (if $p(m) = 0$, suppose passivity of all agents).

In the anonymous model, a Nash equilibrium satisfies the equation [18]

$$F(p) = p, \tag{4.5}$$

where $F(p) = |\{i \in N : m_i < p\}|$ indicates the number of agents whose thresholds are less than p. Evidently, the CBE corresponds to the *minimal* solution of Eq. (4.5).

Thus, one easily calculates the CBE under known thresholds and reputations of the agents. To proceed, let us study control problems. Imagine the influence and/or thresholds of agents can be modified. How should this be done for implementing a required CBE? In terms of the practical interpretations of the model, we aim at reducing the number of agents deciding "to be active."

The aggregated index of mob state is the number of active agents: $K(T, \theta) = |Q_{q(T, \theta)}|$.

In the model with reputation, replace the matrix T with the vector r. In the anonymous case, we have $K(m) = p(m)$.

Let T^0 and θ^0 be the vectors of initial values of influence matrices and agents' thresholds, respectively. Suppose that the following parameters are given: *the admissible sets* of the influences and thresholds of agents (T and Θ, respectively), the *Principal's payoff* $H(K)$ from an achieved mob state K and his *costs* $C(T, \theta, T^0, \theta^0)$ required for modifying the reputations and thresholds of the agents.

As a control efficiency criterion, select the Principal's goal function representing the difference between the payoff $H(\cdot)$ and the costs $C(\cdot)$. Then the *control problem* takes the form

$$H(K(T,\theta)) - C(T,\theta,T^0,\theta^0) \to \max_{T \in \mathrm{T}, \theta \in \Theta}. \tag{4.6}$$

In the anonymous case, the control problem (4.6) becomes

$$H(p(m)) - C(m,m^0) \to \max_{m \in M}, \tag{4.7}$$

where M is the admissible set of threshold vectors in the anonymous case, while m and m^0 designate the terminal and initial threshold vectors, respectively.

Now, consider special cases of the general problem (4.6). The threshold control problem in the anonymous case is treated in Sect. 4.2. And the reputation control problem in the non-anonymous case can be found in Sect. 4.3.

4.2 Threshold Control

Assume that the Principal strives for making the number of active agents equal to a given quantity $K^* \geq 0$. In other words, the Principal aims at *implementing* a new CBE with the number of active agents K^* that does not exceed the old CBE with the number $p(m)$ of active agents. Here a practical interpretation is that the Principal reduces the number of agents acting in a CBE. Controlling the values of thresholds, the Principal must transfer the CBE to a point with the number K^* of active agents. In the anonymous case, the agents have identical reputations, and problem (4.7) takes the form

$$C(m,m^0) \to \min_{m \in \{\eta | p(\eta) = K^*\}}. \tag{4.8}$$

Let g be a nondecreasing function of a nonnegative argument, i.e., the absolute value of the difference between the initial and terminal values of the threshold of a controlled agent. Suppose that the threshold control costs for one agent constitute

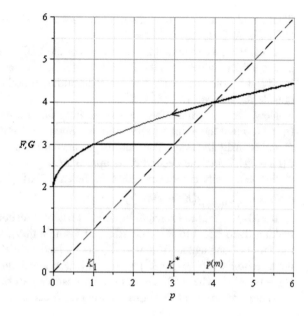

Fig. 4.1 The initial (F) and modified (G) threshold distribution functions

$$c_i(m_i, m_i^0) = g(|m_i - m_i^0|).\qquad(4.9)$$

By implication of costs, $g(0) = 0$. The total costs result from summing up the individual costs:

$$C(m, m^0) = \sum_{i=1}^{n} c_i(m_i, m_i^0) = \sum_{i=1}^{n} g(|m_i - m_i^0|).\qquad(4.10)$$

To elucidate the principle of threshold control, first assume the following. The threshold distribution $F(\cdot)$ is a nondecreasing function defined on the set of non-negative numbers, being left continuous and having the right-hand limit at each point. By analogy with Eq. (4.5), here the equilibrium is the left intersection of the threshold distribution function and the bisecting line of quadrant II. This part of the threshold distribution function can be observed in Fig. 4.1. The thin line shows the threshold distribution function $G(\cdot)$ obtained as the result of control.

Since $p(m)$ is the minimal solution to Eq. (4.5) (see point $(4, 4)$ on Fig. 4.1), then

$$\forall K < p(m) : F(K) > K.\qquad(4.11)$$

Two cases are possible, namely, $F(0+) \le K^*$ and $F(0+) > K^*$ (e.g., $K^* = 3$ and $K^* = 1$, respectively).

If $F(0+) \le K^*$, solve the equation $F(K_1) = K^*$ to find the point $K_1 = F^{-1}(K^*)$.

By virtue of (4.11), $K^* = F(K_1) > K_1 = F^{-1}(K^*)$. The control process consists in that the Principal modifies all thresholds within the interval $[F^{-1}(K^*), K^*)$, making them equal to K^*. The new distribution function $G(\cdot)$ is K^* on the whole interval $[F^{-1}(K^*), K^*]$. Hence, $G(K^*) = K^*$.

Thus, we have obtained the new value of the CBE with the number K^* of active agents, which agrees with the Principal's goals. The new distribution function $G(\cdot)$ in Fig. 4.1 coincides with the old distribution function $F(\cdot)$ at all points lying outside the interval $[F^{-1}(K^*), K^*)$.

Now, let us estimate the costs of shifting the thresholds from a small interval $(t_1, t_2]$ (around the point t) to the point K^*. Suppose that on this interval the function F varies insignificantly. The number of agents on the above interval constitutes $F(t_2) - F(t_1)$, and the desired costs approximate $g(K^* - t)[F(t_2) - F(t_1)]$.

One can easily demonstrate that, in the case $F(0) \leq K^*$, the Principal's control costs are $\int_{F^{-1}(K^*)}^{K^*} g(K^* - t)dF(t)$.

If $F(0+) > K^*$, the Principal modifies the zero thresholds of $[F(0+) - K^*]$ agents, making them equal to K^*. According to formula (4.9), the costs of modifying the threshold of one agent (from 0 to K^*) are defined by $g(K^*)$. The costs of modifying the thresholds of $[F(0+) - K^*]$ agents with zero thresholds can be rewritten as $g(K^*)$ $(F(0+) - K^*)^+$, where $(\cdot)^+$ designates the positive part operator.

Therefore, the total Principal's costs (4.10) become

$$c(K^*) = g(K^*)(F(0+) - K^*)^+ + \int_{F^{-1}(K^*)}^{K^*} g(K^* - t)dF(t). \qquad (4.12)$$

Assertion 4.2 Assume that the Principal implements the CBE with the number K^* of active agents. Then costs (4.12) are minimal required.

Proof of Assertion 4.2. By definition of the function $F(\cdot)$, to reduce the value of $F(K^*)$ (actually, exceeding K^*—see (4.11)), one should proceed as follows. Increase the thresholds lying to the left from the point K^* by a certain quantity such that their values are not smaller than K^*.

As the cost function g is nondecreasing, modifying the values of the thresholds greater than K^* incurs unjustified costs. Therefore, costs minimization dictates that the new value of the thresholds equals K^*.

Consider the set of all possible results of control:

$$\Omega = \left\{ A = \bigcup_{i=1}^{q} [a_i, b_i) | b_q \leq K^*; a_{i+1} > b_i, i = \overline{1, q}; \sum_{i=1}^{q} [F(b_i) - F(a_i)] = F(K^*) - K^* \right\}.$$

Control lies in shifting all thresholds entering these intervals to the point K^*. Evidently, $[F^{-1}(K^*), K^*) \in \Omega$.

The control costs on the set $A \in \Omega$ take the form

$$\int_A g(K^* - t)dF(t) = \sum_{i=1}^{q} \int_{a_i}^{b_i} g(K^* - t)dF(t).$$

Compare the total costs of modifying the thresholds from the interval $[F^{-1}(K^*), K^*)$ and from an arbitrary set $A \in \Omega$. These costs can be expanded into the sum of two terms as follows:

$$\int_{[F^{-1}(K^*),K^*)} g(K^* - t)dF(t) = \int_{A \cap [F^{-1}(K^*),K^*)} g(K^* - t)dF(t) + \int_{[F^{-1}(K^*),K^*) \setminus A} g(K^* - t)dF(t)$$

(4.13)

$$\int_A g(K^* - t)dF(t) = \int_{A \cap [F^{-1}(K^*),K^*)} g(K^* - t)dF(t) + \int_{A \setminus [F^{-1}(K^*),K^*)} g(K^* - t)dF(t).$$

(4.14)

The first summands in equalities (4.13) and (4.14) are identical.

Even if the total costs for the sets A and $[F^{-1}(K^*), K^*)$ differ, this would be observed on the sets $A \setminus [F^{-1}(K^*), K^*)$ and $[F^{-1}(K^*), K^*) \setminus A$ exclusively. Due to definition of the set Ω,

$$\sum_{A \setminus [F^{-1}(K^*),K^*)} [F(b_i) - F(a_i)] = \sum_{[F^{-1}(K^*),K^*) \setminus A} [F(b_i) - F(a_i)].$$

The costs of shifting the thresholds from the sets $A \setminus [F^{-1}(K^*), K^*)$ and $[F^{-1}(K^*), K^*) \setminus A$ can be assigned the lower and upper estimates

$$\int_{A \setminus [F^{-1}(K^*),K^*)} g(K^* - t)dF(t) \geq \min_{t \in A \setminus [F^{-1}(K^*),K^*)} g(K^* - t) \sum_{A \setminus [F^{-1}(K^*),K^*)} [F(b_i) - F(a_i)],$$

(4.15)

$$\int_{[F^{-1}(K^*),K^*) \setminus A} g(K^* - t)dF(t) \leq \max_{t \in [F^{-1}(K^*),K^*) \setminus A} g(K^* - t) \sum_{[F^{-1}(K^*),K^*) \setminus A} [F(b_i) - F(a_i)].$$

(4.16)

By monotonicity of the cost function, we have the inequality

$$\max_{t \in [F^{-1}(K^*),K^*) \setminus A} g(K^* - t) \leq \min_{t \in A \setminus [F^{-1}(K^*),K^*)} g(K^* - t). \qquad (4.17)$$

On the other hand, formulas (4.15)–(4.17) imply

$$\int\limits_{[F^{-1}(K^*),K^*)\setminus A} g(K^* - t)dF(t) \leq \int\limits_{A\setminus[F^{-1}(K^*),K^*)} g(K^* - t)dF(t). \tag{4.18}$$

And expressions (4.13), (4.14) and (4.18) yield

$$\int\limits_{[F^{-1}(K^*),K^*)} g(K^* - t)dF(t) \leq \int\limits_{A} g(K^* - t)dF(t). \bullet$$

Corollary 4.1 *Optimal control modifies only the thresholds belonging to the interval* $[F^{-1}(K^*), K^*)$ *if* K^* *lies within the domain of the function* $F^{-1}(\cdot)$. *In the case when* K^* *is outside the domain of* $F^{-1}(\cdot)$, *optimal control modifies the thresholds belonging to the interval* $[0, K^*)$.

Corollary 4.2 *Solution to the threshold control problem does not explicitly depend on the initial CBE.*

Corollary 4.3 *The CBE obtained by optimal control is unstable.*
 Indeed, a small variation of the thresholds to the left from the point K^* *would "drive" the system to a new equilibrium position (i.e., to the intersection point with bisecting line; e.g., see the point* (4, 4) *in Fig. 4.1). To guarantee stability, it is necessary to shift the thresholds to the right from the point* K^*.

Corollary 4.4 *Assertion 4.2 remains in force when the costs of modifying the threshold of an agent make up* $g(|m_i - m_i^0|)/L(m_i^0)$, *where* $L(\cdot)$ *is indicates any measurable strictly increasing function of the initial threshold.*

Solution (4.12) to problem (4.8) being available, one can revert to problem (4.7). The latter acquires the form of the following scalar optimization problem: $H(K^*) - c(K^*) \to \max_{K^*}$.

In Examples 4.1 and 4.2 below, we present the analytical solution to the problem (4.8) for a series of special cases.

Example 4.1 Consider the relative thresholds $\theta_i = m_i/n$. Assume that their distribution function represents the Pareto distribution with a parameter $\alpha > 1$, i.e.,

$$F_{\alpha\beta}(x) = \frac{x^\alpha + \beta}{1 + \beta}, \quad x \leq 1, \quad \beta > 0. \tag{4.19}$$

Then the density function has the form $f_{\alpha\beta}(x) = \alpha x^{\alpha-1}/(1+\beta)$.

The inverse function to the distribution function is defined by $F_{\alpha\beta}^{-1}(x) = ((1+\beta)x - \beta)^{1/\alpha}$ on the segment $\frac{\beta}{1+\beta} \leq x \leq 1$.

The CBE Eq. (4.5) becomes

$$x^\alpha - (1+\beta)x + \beta = 0. \tag{4.20}$$

Let p be the minimal positive root of Eq. (4.20). Since the distribution function (4.19) increases strictly, we naturally arrive at $p > \beta/(1+\beta)$.

Introduce the cost function $g(x) = |x|$.

Suppose that the goal equilibrium is $k^* = K^*/n$. Consequently, the costs of implementing this CBE, $p \geq k^* \geq \beta/(1+\beta)$, are defined by

$$C_{\alpha\beta}(k^*) = \int_{F_{\alpha\beta}^{-1}(k^*)}^{k^*} g(k^* - t)dF_{\alpha\beta}(t) = k^*\left(F_{\alpha\beta}(k^*) - k^*\right) - \frac{\alpha}{1+\beta}\int_{F_{\alpha\beta}^{-1}(k^*)}^{k^*} t^\alpha dt$$

$$= k^*\left(\frac{(k^*)^\alpha + \beta}{1+\beta} - k^*\right) - \frac{\alpha\left((k^*)^{\alpha+1} - ((1+\beta)k^* - \beta)^{\alpha+1/\alpha}\right)}{(1+\beta)(1+\alpha)}.$$

$$\tag{4.21}$$

Now, assume that $0 < k^* < \beta/(1+\beta)$. Then the costs of implementing this CBE constitute

$$C_{\alpha\beta}(k^*) = k^*\left(\frac{\beta}{1+\beta} - k^*\right) + \int_0^{k^*} g(k^* - t)dF_{\alpha\beta}(t) = k^*\left(\frac{\beta}{1+\beta} - k^*\right)$$

$$+ \frac{(k^*)^{\alpha+1}}{1+\beta} - \frac{\alpha}{1+\beta}\int_0^{k^*} t^\alpha dt = k^*\left(\frac{\beta}{1+\beta} - k^*\right) + \frac{(k^*)^{\alpha+1}}{(1+\beta)(1+\alpha)}.$$

$$\tag{4.22}$$

For $\alpha = 1.3; 1.5; 1.75; 2$ and $\beta = 0.25$, the family of the cost functions (4.21) and (4.22) is illustrated by Fig. 4.2. Clearly, the costs grow as the parameter α of the Pareto distribution goes down. Hence, implementing the goal CBE requires higher relative costs for more uniform distribution of the thresholds (smaller values of the parameter α).

The maximum costs are attained at the points where the distribution function has the largest "distance" to the bisecting line of quadrant I (see Fig. 4.1). This feature results in that the thresholds of relatively many agents should be modified by greater quantities (and the costs increase).

For all values of the parameters α and β, the costs of implementing the zero CBE actually vanish. Indeed, it suffices to shift the zero thresholds by a small quantity to derive a small CBE. Imagine that the Principal strives for implementing a new CBE being smaller than p (the CBE without control). In this case, optimal control lies in modifying the zero thresholds by a small quantity, thereby eliminating the ringleaders from the mob.•

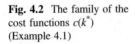

Fig. 4.2 The family of the cost functions $c(k^*)$ (Example 4.1)

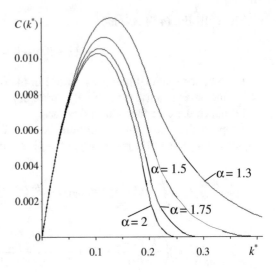

In Example 4.1, the control cost function $g(\cdot)$ does not explicitly depend on the initial values of the agents' thresholds. Instead, it is fully described by the value of threshold variations. In the next example, we analyze another cost function being explicitly dependent on the initial value of the thresholds.

Example 4.2 Within the conditions of Example 4.1, suppose that the cost function of thresholds modification takes the form

$$g(x_1, x_2) = |x_1 - x_2|/x_1. \tag{4.23}$$

Let the goal CBE be k^*. Consequently, the costs for $p \geq k^* \geq \beta/(1+\beta)$ make up

$$C_{\alpha\beta}(k^*) = \int_{F_{\alpha\beta}^{-1}(k^*)}^{k^*} g(k^*, k^* - t) dF_{\alpha\beta}(t) = \left(F_{\alpha\beta}(k^*) - k^*\right) - \frac{\alpha}{(1+\beta)k^*} \int_{F_{\alpha\beta}^{-1}(k^*)}^{k^*} t t^{\alpha-1} dt$$

$$= \left(\frac{(k^*)^\alpha + \beta}{1+\beta} - k^*\right) - \frac{\alpha\left((k^*)^{\alpha+1} - ((1+\beta)k^* - \beta)^{\alpha+1/\alpha}\right)}{k^*(1+\beta)(1+\alpha)}.$$

Now, assume that $0 < k^* < \beta/(1+\beta)$. On this interval, the costs turn out infinitely large: we should modify the zero thresholds by the quantity k^*, which requires infinitely large costs according to formula (4.23). And so, the Principal can implement only a CBE from the interval $p \geq k^* \geq \beta/(1+\beta)$. For $\alpha = 1.3; 1.5; 1.75;$ 2 and $\beta = 0.25$, the family of the cost functions can be observed in Fig. 4.3.•

In Example 4.2, the control cost function possesses the monotonous property, which is easily explained: the larger is the deviation of the new CBE (implemented

Fig. 4.3 The family of the cost functions $c(k^*)$ (Example 4.2)

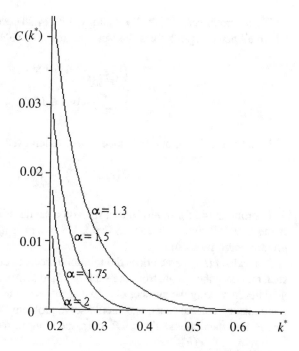

by the Principal during threshold control) from the old one (without such control), the higher are the corresponding costs.

4.3 Reputation Control

Let the non-anonymous case take place and the agents' thresholds be fixed. Consider the reputation control problem. Suppose that the Principal seeks for making the number of active agents not greater than a given quantity $K^* \geq 0$. In this case, the control problem acquires the form

$$C\left(r, \theta, r^0, \theta\right) \to \min_{r \in R \cap \left\{\eta \mid |Q_{q(\eta, \theta^0)}| \leq K^*\right\}}.$$

In the initial CBE, all active agents form the set Q_k. In other words, we have the following system of inequalities:

$$\begin{cases} \sum\limits_{j \in Q_k \setminus \{i\}} r_j^0 \geq \theta_i, i \in Q_k, \\ \sum\limits_{j \in Q_k} r_j^0 < \theta_i, i \in N \setminus Q_k. \end{cases} \tag{4.24}$$

Fix a certain set $P \subseteq N$. By analogy with (4.24), express the conditions under which all active agents form this set with the new values of reputations:

$$
\begin{cases}
\sum_{j \in P \setminus \{i\}} r_j \geq \theta_i, i \in P, \\
\sum_{j \in P} r_j < \theta_i, i \in N \setminus P.
\end{cases}
$$

Denote by $c(P)$ the optimal value of a criterion used in the optimization problem

$$
C(r, \theta, r^0, \theta) \to \min_{r:(24)} . \tag{4.25}
$$

In problem (4.25), minimization runs on a set described by n linear inequalities. If the cost function is linear (convex), we obtain a linear (convex, respectively) programming problem.

The value $c(P)$ characterizes the minimum costs to control the agents' reputation such that only the agents from the set $P \subseteq N$ are active. Imagine that the Principal's goal lies in making the number of active agents equal to a given quantity $K^* \geq 0$. To minimize the control costs, one should solve problem (4.25) for each of K^*-element sets P, and then choose the set P^* corresponding to the minimum costs: $P^* = \arg \min_{P \in \{W \in 2^N : |W| = K^*\}} c(P)$.

4.4 Reflexive Control

We have studied two cases of Principal's control, namely, the impacts exerted on the agents' thresholds and on the agents' reputation. Now, let us analyze the capabilities of *reflexive control*—the Principal influences the beliefs of the agents about their parameters, the beliefs about beliefs, etc. [74, 75]. Select the agents' thresholds as the subject of control. *Reflexive control* forms the following awareness structures of the agents: θ_{ij}—the beliefs of agent i about the threshold of agent j (an awareness structure of rank 2 or depth 2); θ_{ijk}—the beliefs of agent i about the beliefs of agent j about the threshold of agent k (an awareness structure of rank 3 or depth 3), and so on. Possessing a certain awareness structure, the agents choose their actions as an *informational equilibrium* [75]. Notably, each agent chooses a specific action as the best response to the actions expected by him from the opponents (according to his awareness structure).

Recall the results derived in the previous subsections that characterize the thresholds leading to a desired CBE. For convenience, we believe that any result achievable via a real variation of thresholds can be implemented by informational control (an appropriate modification of the agents' beliefs about their thresholds). And so, informational control of the thresholds turns out equivalent to threshold control in a common sense. Apparently, the former type of control is "softer" than the latter.

Nevertheless, informational control implementation in mob control problems faces an obstacle. One property of "good" informational control concerns its *stability* [75] when all agents observe the results they actually expected.

Within the suggested mob model, assume that each agent a posteriori observes the number of agents decided to "be active." (In fact, this is a rather weak assumption in comparison with the mutual observability of the individual actions.) Then, under a stable informational control, each agent observes the number of active agents he actually expects. Stability is a substantial requirement for a long-term interaction between the Principal and agents. Indeed, under an unstable informational control, the agents just once doubting the truth of Principal's information have good reason to do it later.

Assertion 4.3 In the anonymous case, there exists no stable informational equilibrium such that the number of active agents is strictly smaller than in a CBE.

Proof of Assertion 4.3. Denote by Q_Σ the set of agents that act in a stable informational equilibrium. Suppose that their number does not exceed the number of agents acting in a CBE: $|Q_\Sigma| \leq |Q_{p(\theta)}|$. Due to stability of the informational equilibrium, each agent $i \in Q_\Sigma$ satisfies the condition $|Q_\Sigma| - 1 = \sum_{j \neq i} x_j \geq (n-1)\theta_i$. Hence, $|Q_{p(\theta)}| - 1 \geq (n-1)\theta_i$, which implies $i \in Q_{p(\theta)}$. Therefore, $Q_\Sigma \subseteq Q_{p(\theta)}$. If passive agents exist in $Q_{p(\theta)} \backslash Q_\Sigma$, this equilibrium appears unstable for them. And so, $Q_\Sigma = Q_{p(\theta)}$ for the stable informational equilibrium.•

The "negative" result of Assertion 4.3 witnesses to the complexity of implementing long-term informational control of threshold behavior in a mob.

In the current section, we have formulated and solved control problems for collective threshold behavior of a mob in several relevant cases. This has been done by control of the agents' thresholds, agents' reputations and reflexive control. A promising direction of further research is to analyze the efficiency of these types of control applied simultaneously.

Chapter 5
Stochastic Models of Mob Control

This section explores the following model of agents' threshold behavior. Making binary decisions (choosing between "activity" and "passivity"), the agents consider the choice of other members in a group. We formulate and solve an associated control problem, i.e., the random choice problem for the initial states of some agents in order to vary the number of agents preferring "activity" in an equilibrium [25].

Further exposition focuses on *stochastic models* of threshold behavior control. The sets of agents whose thresholds undergo changes or the values of agents' thresholds are chosen randomly (also see Sect. 5.3 and [24]). In practice, a possible tool of such control consists in mass media [48, 75] or any other *unified* (informational, motivational and/or institutional [74]) impacts on agents.

For instance, consider the following interpretations of potential control actions: the thresholds of a given proportion of randomly chosen agents are nullified (which matches "*excitation*") or maximized (which matches "*immunization*," i.e., complete insusceptibility to social pressure). Or just believe that each agent can be excited or/and immunized with a given probability. Such transformations of agents' thresholds cause a corresponding variation in the equilibrium state of an active network structure (a social network, a mob). Readers will find the details below.

Another way for managing threshold behavior (not agents' thresholds) is *staff control* according to the control types classification introduced in [74]. Staff control implies embedding additional agents with zero and maximum thresholds (*provokers* and *immunizers*, respectively) in an ANS. In this case, the resulting equilibrium of the ANS depends on the number of embedded agents having an appropriate type.

And finally, there may exist two control authorities (*Principals*) exerting opposite informational impacts on the agents. This situation of *distributed control* [43, 74] can be interpreted as *informational confrontation* [46, 48, 70, 72] between the Principals. Using the analysis results of the control problems for each separate Principal, one can describe their interaction in terms of game theory.

First, consider agents' threshold control that affects the distribution function of their thresholds, causing mob "excitation."

© Springer International Publishing AG 2017
V.V. Breer et al., *Mob Control: Models of Threshold Collective Behavior*,
Studies in Systems, Decision and Control 85,
DOI 10.1007/978-3-319-51865-7_5

5.1 Mob "Excitation" Control

Let the agents' thresholds be independent identically distributed random variables with a theoretical distribution function $F(\cdot)$. According to expressions (2.7)–(2.9), the CBE of a mob is predetermined by the distribution function of the agents' thresholds. Hence, a control action affecting this distribution function naturally modifies the CBE. Below we study some possible statements of such control problems.

Suppose that a control action nullifies the threshold of each agent independently from the other agents with a same probability $\alpha \in [0, 1]$ for all agents. This model will be called **Model I**. According to (2.1), the agents having zero thresholds choose unit actions regardless of the actions of the others. Thus and so, the parameter α plays the role of the proportion of initially *excited* agents.

Assertion 5.1 After "excitation," the agents' thresholds are described by the distribution function

$$F_\alpha(x) = \alpha + (1-\alpha)F(x). \tag{5.1}$$

Proof of Assertion 5.1. Consider a given vector $\theta = (\theta_1, \ldots, \theta_n)$ of independent identically distributed random variables obeying the same distribution $F(\cdot)$. This vector undergoes the following transformation with the random component: each threshold θ_i is nullified with the probability α. Such transformation yields another random vector θ' with some distribution $F_\alpha(\cdot)$ to-be-found.

Reexpress the components of the vector θ' in the form $\theta_i' = \theta_i \zeta_i$, where $P(\zeta_i = 0) = \alpha$, $P(\zeta_i = 1) = 1 - \alpha$, and all elements of the set $\{\zeta_i, \theta_i\}$ are pairwise independent.

Construct the distribution function $F_\alpha(\cdot)$ of the random variable θ_i':

$$F_\alpha(x) = P(\theta_i' \le x) = P(\theta_i \varsigma_i \le x) = P(\varsigma_i = 0) + P(\varsigma_i = 1, \theta_i \le x).$$

Owing to the independence of ζ_i and θ_i, we have

$$P(\varsigma_i = 1, \theta_i \le x) = P(\varsigma_i = 1)P(\theta_i \le x) = (1 - \alpha)F(x),$$

which brings directly to formula (5.1).•

By substituting the new distribution function (5.1) into Eq. (2.8), evaluate α implementing the CBE y:

$$\alpha(y) = \frac{y - F(y)}{1 - F(y)}. \tag{5.2}$$

If $\alpha(y) < 0$ for some $y \in (0, 1]$, then this CBE cannot be implemented by the control actions under consideration, see expression (5.2).

Denote by $x^*(\alpha)$ the CBE (2.9) corresponding to the distribution function (5.1) and by $W_\alpha = \bigcup_{\alpha \in [0;1]} x^*(\alpha)$ the *attainability set* (i.e., the set of all proportions of agents whose excitation can be implemented as the CBE under a certain control action).

A rather simple analytical form of the distribution function (5.1) gives answers to a series of practical questions.

Assertion 5.2 If $F(\cdot)$ is a strictly convex function such that $F(0) = 0$, then $W_\alpha = [0, 1]$. In other words, any proportion of excited agents can be implemented as the CBE by an appropriate value of the parameter α.

Proof of Assertion 5.2. By virtue of the condition $F(0) = 0$, the boundaries of the unit segment are implemented as the CBE under $\alpha = 0$ and $\alpha = 1$.

Fix an arbitrary point $x_1 \in (0, 1)$. The whole curve of the convex function $F(\cdot)$ lies not higher than the bisecting line and the equation $F_\alpha(x_1) = x_1$ has a solution with $0 \leq \alpha(x_1) < 1$. This fact and formula (5.2) lead to $F_\alpha(x_1) = \alpha(x_1) + (1 - \alpha(x_1))F(x_1) = x_1$ and hence $F(x_1) = \frac{x_1 - \alpha(x_1)}{1 - \alpha(x_1)}$. Consequently, $F'_\alpha(x_1 - 0) = (1 - \alpha(x_1))F'(x_1 - 0)$.

On the other hand, strict monotonicity of the derivative of a strictly convex function dictates that

$$F'(x_1 - 0) = \frac{F'(x_1 - 0)\int_{x_1}^1 dx}{1 - x_1} < \frac{F'(x_1 + 0)\int_{x_1}^1 dx}{1 - x_1} < \frac{\int_{x_1}^1 F'(x + 0)dx}{1 - x_1} = \frac{1 - F(x_1)}{1 - x_1}.$$

Next, we obtain

$$F'_\alpha(x_1 - 0) = (1 - \alpha(x_1))F'(x_1 - 0) < (1 - \alpha(x_1))\frac{1 - \frac{x_1 - \alpha(x_1)}{1 - \alpha(x_1)}}{1 - x_1} < 1.$$

This means that the equilibrium x_1 is stable, since the curve $F(\cdot)$ crosses the bisecting line by approaching it "from left and top."•

Imagine that we know the Principal's payoff $H(x)$ from exciting the proportion of agents x and his control costs $c_\alpha(\alpha)$. In this case, the *mob excitation control problem* admits the following statement:

$$H(x^*(\alpha)) - c_\alpha(\alpha) \to \max_{\alpha \in [0;1]} . \tag{5.3}$$

Example 5.1 Consider several distribution functions:

(I) $F_I(x) = x$,
(II) $F_{II}(x) = x^2$,
(III) $F_{III}(x) = \sqrt{x}$.

For the distribution functions (I)–(III), formula (5.2) gives:

- $\alpha^I(y) = 0$, i.e., a unique CBE is $W^I = \{1\}$;
- $\alpha^{II}(y) = \frac{y}{1+y}$, $x^{II*}(\alpha) = \frac{1-|1-2\alpha|}{2(1-\alpha)}$, $W_\alpha^{II} = [0, 1]$;
- $\alpha^{III}(y) = -\sqrt{y} \leq 0$, i.e., a unique CBE is again $W_\alpha^{III} = \{1\}$.•

A possible "dynamic" generalization of the above model is when at each step t an agent can be independently excited with the probability α (including "multiple" excitation of a same agent). In this case, we have the distribution function

$$F_\alpha(t,x) = 1 - (1-\alpha)^t + (1-\alpha)^t F(x), \quad t = 0, 1, 2, \ldots. \tag{5.4}$$

Another "dynamic" generalization–different excitation probabilities at different steps–is reducible to the single-step case. Really, the distribution function corresponding to two steps with independent excitation probabilities α_1 and α_2 of independent excitation possesses form (5.1) where

$$\alpha = \alpha_1 + \alpha_2 - \alpha_1\alpha_2. \tag{5.5}$$

Now, analyze an alternative approach to equilibrium control (**Model II**). Here the set N (recall that $\#N = n$) is supplemented by k external *provokers* (the set K). They have the thresholds $\theta_i = 0 \ \forall \ i \in K$, i.e., always act. Then the probability that an arbitrarily chosen agent from the new set $N \cup K$ has a threshold not exceeding x comprises the probabilities of two independent events, namely:

(1) the probability that the chosen agent represents an external provoker:

$$\frac{k}{k+n};$$

(2) the probability that the chosen agent is not an external provoker and his threshold does not exceed x:

$$\left(1 - \frac{k}{k+n}\right) F(x).$$

Consequently, we have constructed a new set of agents whose thresholds form independent identically distributed random variables with the distribution function

$$F_K(x) = \frac{k}{k+n} + \left(1 - \frac{k}{k+n}\right) F(x). \tag{5.6}$$

Within the framework of Model I, the parameter α (the excitation probability of an arbitrary agent) can be also comprehended as the probability of meeting such an agent with zero threshold. Therefore, it seems reasonable to introduce the same notation for the probability of meeting an external provoker in Model II:

$$\alpha = \frac{k}{k+n}.$$

Direct comparison of the distribution functions (5.1) and (5.6) in the case of "exciting" a proportion of agents testifies to the equivalence of Models I and II.

The next subsection considers "inverse" control which aims at reducing excitation (the so-called mob "immunization").

5.2 Mob "Immunization" Control

Suppose that, as the result of a control action, the threshold of each agent becomes 1 independently from the other agents with a same probability $\beta \in [0, 1]$ for all agents. According to (2.1), the agents having unit thresholds are passive, and the parameter β plays the role of the proportion of initially "*immunized*" agents. By analogy with Assertion 5.1, one can show that after "immunization" the agents' thresholds are described by the distribution function

$$F_\beta(x) = \begin{cases} (1-\beta)F(x), x \in [0;1), \\ 1, x = 1. \end{cases} \tag{5.7}$$

Denote by $x^*(\beta)$ the CBE corresponding to the distribution function (5.7) and by $W_\beta = \bigcup_{\beta \in [0;1]} x^*(\beta)$ the attainability set.

By substituting the new distribution function (5.7) into Eq. (2.8), we can evaluate β implementing the CBE y:

$$\beta(y) = 1 - \frac{y}{F(y)}. \tag{5.8}$$

If $\beta(y) < 0$ for some $y \in (0, 1]$, then this CBE cannot be implemented by the control actions under consideration, see expression (5.8).

Assertion 5.2a If $F(\cdot)$ is a strictly concave function such that $F(0) = 0$, then $W_\beta = [0, 1]$. In other words, any proportion of immunized agents can be implemented as the CBE by an appropriate value of β.

This result can be proved just like Assertion 5.2.

Example 5.2 For the distribution functions (I)–(III), formula (5.8) gives:

- $\beta^I(y) = 0$, i.e., a unique CBE is $W_\beta^I = \{0\}$;
- $\beta^{II}(y) = 1 - \frac{1}{y} \le 0$, i.e., a unique CBE is again $W_\beta^{II} = \{0\}$;
- $\beta^{III}(y) = 1 - \sqrt{y}$, $x^{III*}(\beta) = (1-\beta)^2$, $W_\beta^{III} = [0, 1]$.•

Example 5.3 Solve the control problem (5.3) with the distribution function (III). The Principal seeks for minimizing the proportion of the excited agents: $H(x) = -x$. His control costs are described by $c_\beta(\beta) = \lambda \beta$, where $\gamma \geq 0$. This statement brings to the optimization problem $-(1 - \beta)^2 - \lambda\beta \rightarrow \max_{\beta \in [0;1]}$. The solution $\beta^* = 1 - \frac{\lambda}{2}$ corresponds to implementation of the CBE $\frac{\lambda^2}{4}$. •

Similarly to the previous subsection, consider Model II—an alternative approach to equilibrium control where the set N is supplemented by l external *immunizers* (the set L). They have the thresholds $\theta_i = 1 \; \forall \; i \in L$ and never act. Then the probability that an arbitrarily chosen agent from the new set $N \cup K$ has a threshold not exceeding $x < 1$ coincides with the probability that the chosen agent is not an immunizer and his threshold does not exceed $x < 1$:

$$\left(1 - \frac{l}{l+n}\right)F(x).$$

The threshold of an arbitrarily chosen agent is surely not greater than 1. Thus and so, we have constructed the new set $N \cup L$ of agents whose thresholds form independent identically distributed random variables with the distribution function

$$F_L(x) = \begin{cases} \left(1 - \frac{l}{l+n}\right)F(x), & x < 1, \\ 1, & x = 1. \end{cases} \tag{5.9}$$

In Model I, the parameter β (the immunization probability of an arbitrary agent) can be also viewed as the probability of meeting such an agent with unit threshold. Therefore, we introduce the same notation for the probability of meeting an external immunizer in Model II:

$$\beta = \frac{l}{l+n}.$$

Direct comparison of the distribution functions (5.7) and (5.9) in the case of "immunizing" a proportion of agents testifies to the equivalence of Models I and II.

Now, we study simultaneous control of two Principals, one of them concerned with mob excitation and the other with its immunization.

5.3 Informational Confrontation

In the previous considerations, we have treated a mob as the object of control actions chosen by a single subject (*Principal*). There may exist several subjects interested in certain states of a network and applying control actions to it (the so-called *distributed control system* [59, 74]). In this case, the control subjects interact via informational impacts exerted by them on the network, which leads to their *informational confrontation* (see the survey [48]). Generally, such situations are described by a normal-form game of the Principals, and their strategies predetermine the parameters of the agents' game [74]. For instance, take the models of informational confrontation in social networks [46, 72], on cognitive maps [55, 68] and others discussed in the overview [76]. As it was outlined in [73], more sophisticated situations are also possible when control actions appear "unsymmetrical" (e.g., in the "attack/defense" situation, Principal 1 affects the initial states of agents, whereas Principal 2 modifies the structure of relations among them or/and their thresholds; Principal 2 can act simultaneously with the opponent or immediately after the latter, being aware of his choice). Such situations are described using models of hierarchical games.

Let us investigate informational confrontation in the following setting. There are Principal 1 "exciting" a proportion $\alpha \in [0, 1]$ of agents and Principal 2 "immunizing" a proportion $\beta \in [0, 1]$ of agents (alternatively, each agent can be independently excited or/and immunized with a corresponding probability). For definiteness, suppose that if a certain agent is excited and immunized simultaneously, then his threshold remains the same (Model I). We obtain the following distribution function of the agents' thresholds:

$$F_{\alpha,\beta}(x) = \begin{cases} \alpha(1 - \beta) + (1 - \alpha - \beta + 2\alpha\beta)F(x), x \in [0; 1), \\ 1, x = 1. \end{cases} \tag{5.10}$$

Denote by $x^*(\alpha, \beta)$ the CBE corresponding to the distribution function (5.10). In addition, let $y(\alpha, \beta) = \inf\{x : x \in (0, 1], F_{\alpha,\beta}(x) = x\}$ signify the least nonzero root of the equation $F_{\alpha,\beta}(x) = x$. According to (2.7) and (2.9), the collective behavior equilibrium is

$$x^*(\alpha, \beta) = \begin{cases} y(\alpha, \beta), \text{if } \forall z \in [0, y(a, \beta)]F_{\alpha,\beta}(z) \geq z, \\ 0, \quad \text{otherwise.} \end{cases}$$

Assertion 5.3 For any $\beta \in [0, 1]$, $x^*(\alpha, \beta)$ is a monotonically nondecreasing function of α.

Proof of Assertion 5.3. Consider the partial derivative $\frac{\partial}{\partial\alpha}F_{\alpha,\beta}(x)$ at an arbitrary point $x \in [0, 1)$:

$$\frac{\partial}{\partial\alpha}F_{\alpha,\beta}(x) = \frac{\partial}{\partial\alpha}(\alpha(1 - \beta) + (1 - \alpha - \beta + 2\alpha\beta)F(x))$$
$$= (1 - \beta) + (2\beta - 1)F(x) = (1 - \beta)(1 - F(x)) + \beta F(x).$$

Note that $(1 - \beta)(1 - F(x)) \geq 0$ and $\beta F(x) \geq 0$. Then we have $\frac{\partial}{\partial\alpha}F_{\alpha,\beta}(x) \geq 0$. Hence, it follows that $F_{\alpha_2,\beta}(x) \geq F_{\alpha_1,\beta}(x)$ under $\alpha_2 > \alpha_1$.

In combination with this result, the fact that $F_{\alpha_1,\beta}(x) \geq x \,\forall x : x < x^*(\alpha_1, \beta)$ under $\alpha_2 > \alpha_1$ brings to the inequality $F_{\alpha_2,\beta}(x) \geq x \,\forall x : x < x^*(\alpha_1, \beta)$. Therefore, $x^*(\alpha_2, \beta) \geq x^*(\alpha_1, \beta).\bullet$

Assertion 5.4 For any $\alpha \in [0, 1]$, $x^*(\alpha, \beta)$ is a monotonically nonincreasing function of β.

Proof of Assertion 5.4 is similar to that of Assertion 5.3.

Designate by $W_{\alpha,\beta} = \bigcup_{(\alpha,\beta)\in[0;1]^2} x^*(\alpha, \beta)$ the *attainability set*.

Let $V_{\alpha,\beta}$ be the set of *attainable equilibria*, i.e., the set of all points becoming stable equilibria (CBE) under a certain control action (α, β). The definition implies $W_{\alpha,\beta} \subseteq V_{\alpha,\beta}$. The points of the set $V_{\alpha,\beta}$ implement the CBE under some choice of x_0 (generally, not coinciding with $x_0 = 0$ accepted in this paper).

By substituting the new distribution function (5.10) into Eq. (2.7), we find the pairs (α, β) implementing the given CBE.

Example 5.4 For the distribution function (I), we obtain

$$x^{I*}(\alpha, \beta) = \frac{\alpha(1 - \beta)}{\alpha + \beta - 2\alpha\beta}. \tag{5.11}$$

The curve of the distribution function $F^1_{\alpha,\beta}(x)$ is illustrated by Fig. 5.1.

In the current example, $W_{\alpha,\beta} \subseteq V_{\alpha,\beta} = [0, 1].\bullet$

To construct the attainability set and identify the class of functions satisfying $W_{\alpha,\beta} = [0, 1]$, rewrite the resulting distribution function of the agents' thresholds in the form (Fig. 5.2).

$$F_{\alpha,\beta}(x) = \begin{cases} \delta(\alpha, \beta) + k(\alpha, \beta)F(x), & x \in [0, 1), \\ 1, & x = 1, \end{cases}$$

where

$$\begin{cases} \delta(\alpha, \beta) = \alpha(1 - \beta), \\ k(\alpha, \beta) = 1 - \alpha - \beta + 2\alpha\beta. \end{cases} \tag{5.12}$$

Fig. 5.1 $x^{I*}(\alpha, \beta)$ under $F(x) = x$

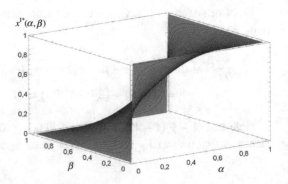

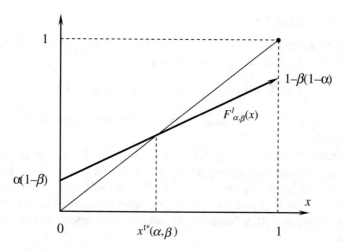

Fig. 5.2 The curve of the distribution function $F^l_{\alpha,\beta}(x)$

Obviously, far from any values $\delta \in [0, 1]$, $k \in [0, 1]$ can be obtained via applying transformation (5.12) to some $\alpha \in [0, 1]$, $\beta \in [0, 1]$. Moreover, the constraint $k + \delta \leq 1$ following from the properties of the distribution function is not unique, see Fig. 5.3.

Assertion 5.5 The value domain of transformation (5.12) represents the set of points (δ, k) in the unit square $[0, 1] \times [0, 1]$ satisfying the conditions

$$k \leq 1 - \delta, \quad k \geq 2\sqrt{\delta}\left(1 - \sqrt{\delta}\right). \tag{5.13}$$

Fig. 5.3 The value domain of the transformation $(\alpha, \beta) \rightarrow (\delta, k)$ (*grey color*)

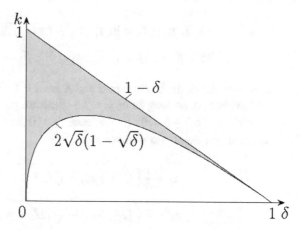

Proof of Assertion 5.5

1. We demonstrate that $k(\alpha, \beta) \leq 1 - \delta(\alpha, \beta)$ for $\alpha \in [0, 1]$, $\beta \in [0, 1]$. Substitute $\delta(\alpha, \beta)$ into formula (5.12) of $k(\alpha, \beta)$ to get

$$k(\alpha, \beta) = 1 - \alpha - \beta + 2\alpha\beta = 1 - \delta(\alpha, \beta) - \beta(1 - \alpha).$$

Having in mind the inequality $\beta(1 - \alpha) \geq 0$, we establish that $k(\alpha, \beta) \leq 1 - \delta(\alpha, \beta)$.

2. Now, prove that $k(\alpha, \beta) \geq 2\sqrt{\delta(\alpha, \beta)}\left(1 - \sqrt{\delta(\alpha, \beta)}\right)$ for $\alpha \in [0, 1]$, $\beta \in [0, 1]$. If $\alpha = 0$, the inequality clearly holds, as $\delta(\alpha, \beta) = 0$ and $k(\alpha, \beta) = 1 - \beta \geq 0$. Under $\beta = 0$, by analogy we obtain $\delta(\alpha, \beta) = 0$ and $k(\alpha, \beta) = 1 - \alpha \geq 0$.
In the case $\alpha > 0$, $\beta > 0$, use the first equation of system (5.12) to express β through α and $\delta(\alpha, \beta)$, substituting the result into the second equation of system (5.12):

$$k(\alpha, \beta) = 1 - \delta(\alpha, \beta) - \left(1 - \frac{\delta(\alpha, \beta)}{\alpha}\right)(1 - \alpha) = \alpha + \frac{\delta(\alpha, \beta)}{\alpha} - 2\delta(\alpha, \beta).$$

$$(5.14)$$

Fix $\delta = \delta(\alpha, \beta) = const$ and minimize $k(a, \delta)$ in α. The first-order necessary optimality condition has the form $\frac{\partial k(\alpha, \delta)}{\partial \alpha} = 1 - \frac{\delta}{\alpha^2}$. The equation $1 - \delta/\alpha^2 = 0$ possesses the unique positive solution $\alpha = \sqrt{\delta}$ for $\delta \in (0, 1]$ (which is the case under $\alpha > 0$, $\beta > 0$). The function $k(\alpha, \delta)$ reaches its minimum in α at the point $\alpha = \sqrt{\delta}$: $\min_{\alpha \in (0,1]} k(\alpha, \delta) = k\left(\sqrt{\delta}, \delta\right) = 2\sqrt{\delta}(1 - \delta)$.
It remains to demonstrate that any point (δ_0, k_0) belonging to this domain is an image of some point of the unit square.

3. Let us prove that

$$\forall \delta_0 \in [0, 1], k_0 \in [0, 1] : 2\sqrt{\delta_0}\left(1 - \sqrt{\delta_0}\right) \leq k_0 \leq 1 - \delta_0$$

$$\exists \alpha \in [0, 1], \exists \beta \in [0, 1] : \delta_0 = \delta(\alpha, \beta), k_0 = k(\alpha, \beta).$$

If $\delta_0 = 0$, the desired values of α and β are $\alpha = 0$ and $\beta = 1 - k_0$. For $\delta_0 > 0$, we have $\alpha > 0$, so long as $\delta_0 = \alpha(1 - \beta)$. Multiply formula (5.14) by α to get $\alpha^2 - \alpha(2\delta_0 + k_0) + \delta = 0$. In the case $\delta_0 > 0$ and $2\sqrt{\delta_0}(1 - \sqrt{\delta_0}) \leq k_0$, this equation possesses two positive roots:

$$\alpha_1 = \frac{1}{2}\left((2\delta_0 + k_0) + \sqrt{(2\delta_0 + k_0)^2 - 4\delta_0}\right),$$

$$\alpha_2 = \frac{1}{2}\left((2\delta_0 + k_0) - \sqrt{(2\delta_0 + k_0)^2 - 4\delta_0}\right).$$

Fig. 5.4 Design procedure of attainability set: an illustration

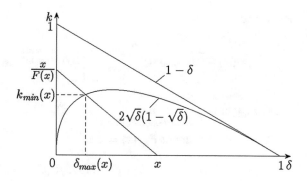

Moreover, $\alpha_1 \in [0, 1]$ and $\alpha_2 \in [0, 1]$ provided that $k_0 \leq 1 - \delta_0$. By choosing $\beta_1 = 1 - \frac{\delta_0}{\alpha_1} = 1 - \alpha_2$ and $\beta_2 = 1 - \frac{\delta_0}{\alpha_2} = 1 - \alpha_1$,   we   finally   arrive   at $\delta_0 = \delta(\alpha_1, \beta_1)$, $k_0 = \delta(\alpha_2, \beta_2)$.•

The attainability set can be constructed for an arbitrary smooth distribution function $F(x)$ using the following statement (see Fig. 5.4).

Assertion 5.6 A point $x \in [0, 1]$ belongs to the set of attainable equilibria of a distribution function $F(\cdot) \in C[0, 1]$ if and only if either $F(x) = 0$, or

$$\frac{F'(x)}{F(x)} \cdot \left(x - \frac{\left(\sqrt{1 + \frac{x}{F(x)} \left(\frac{1}{F(x)} - 2 \right)} - 1 \right)^2}{\left(\frac{1}{F(x)} - 2 \right)^2} \right) < 1. \tag{5.15}$$

(If $F(x) = 1/2$, the parenthesized expression should be comprehended as an appropriate finite limit, and condition (5.15) takes the form $2\, x\, (1 - x)\, F' < 1$.)

Proof of Assertion 5.6. Recall that the belonging of a point to the attainability set means the existence of a pair $(\alpha, \beta) \in [0, 1] \times [0, 1]$ such that $F_{\alpha,\beta}(x) = x$ (the equilibrium condition) and $F'_{\alpha,\beta}(x) < 1$ (the equilibrium stability condition), where $F_{\alpha,\beta}(x)$ is transformation (5.10) of the function $F(x)$. In other words, a point belongs to the attainability set if and only if

$$\min_{\{\alpha \in [0,1], \beta \in [0,1]: F_{\alpha,\beta}(x) = x\}} F'_{\alpha,\beta}(x) < 1$$

and the set $\{\alpha \in [0, 1], \beta \in [0, 1] : F_{\alpha,\beta}(x) = x\}$ appears nonempty.

The derivative $F'_{\alpha,\beta}(x)$ has the form

$$F'_{\alpha,\beta}(x) = k(\alpha, \beta) F'(x)$$

and hence

$$\min_{\left\{\alpha\in[0,1],\,\beta\in[0,1]:F_{\alpha,\beta}(x)=x\right\}} F'_{\alpha,\beta}(x) = \min_{\left\{\alpha\in[0,1],\,\beta\in[0,1]:F_{\alpha,\beta}(x)=x\right\}} k(\alpha,\beta)\cdot F'(x)$$
$$= k_{\min}(x)F'(x),$$

where $k_{\min}(x) = \min_{\left\{\alpha\in[0,1],\,\beta\in[0,1]:F_{\alpha,\beta}(x)=x\right\}} k(\alpha,\beta).$

Rewrite the constraint $F_{\alpha,\beta}(x) = x$ as

$$\delta(\alpha,\beta) + k(\alpha,\beta)F(x) = x.$$

Consider separately the case $F(x) = 0$. The smoothness of the distribution function dictates $F'_{\alpha,\beta}(x) = F'(x) = 0$. Furthermore, the constraint is satisfied under $\alpha = x$, $\beta = 0$, since $\delta(x,0) = x$. This means that, if $F(x) = 0$, the point x belongs to the attainability set.

For $F(x) > 0$, draw the constraint on the plane $\delta\, k$ (see Fig. 5.4). Actually, it represents a straight line passing connecting the points $\left(0, \frac{x}{F(x)}\right)$ and $(x,0)$.

Clearly, for any x and $F(x)$, the minimum value $k_{\min}(x)$ is achieved at the intersection point of this line and the curve $2\sqrt{\delta}\left(1 - \sqrt{\delta}\right)$. Denote by $\delta_{\max}(x)$ the abscissa of the intersection point. Then we have

$$\frac{1}{F(x)}\left(\delta_{\max}(x) - x\right) = 2\sqrt{\delta_{\max}(x)}\left(1 - \sqrt{\delta_{\max}(x)}\right).$$

For any x and $F(x)$, this equation possesses a unique root within the segment $[0, 1]$, i.e.,

$$\delta_{\max}(x) = \frac{\left(\sqrt{1 + \frac{x}{F(x)}\left(\frac{1}{F(x)} - 2\right)} - 1\right)^2}{\left(\frac{1}{F(x)} - 2\right)^2}.$$

By evaluating $k_{\min}(x) = \frac{1}{F(x)}(x - \delta_{\max}(x))$ and expressing $F'_{\alpha,\beta}(x)$ through this minimum value, we naturally arrive at formula (5.15).•

Assertion 5.7 A point $x \in [0, 1]$ belongs to the attainability set of a distribution function $F(\cdot) \in C[0, 1]$ if and only if it belongs to its set of attainable equilibria and

$$\min_{y\in[0,x]} \left(\delta_{\max}(x) + k_{\min}(x)F(y) - y\right) \geq 0, \tag{5.6}$$

where

$$\delta_{\max}(x) = \frac{\left(\sqrt{1 + \frac{x}{F(x)}\left(\frac{1}{F(x)} - 2\right)} - 1\right)^2}{\left(\frac{1}{F(x)} - 2\right)^2}, \quad k_{\min}(x) = \frac{1}{F(x)}(x - \delta_{\max}(x)).$$

Proof of Assertion 5.7. Recall that the belonging of a point to the attainability set means the existence of a pair $(\alpha, \beta) \in [0, 1] \times [0, 1]$ such that $F_{\alpha,\beta}(x) = x$ (the equilibrium condition) and $F'_{\alpha,\beta}(x) < 1$ (the equilibrium stability condition), and $\min_{y \in [0,x]} \left(F_{\alpha,\beta}(y) - y\right) \geq 0$, where $F_{\alpha,\beta}(x)$ is transformation (5.10) of the function $F(x)$.

We have earlier demonstrated that, under $\delta(\alpha, \beta) = \delta_{\max}(x)$ and $k(\alpha, \beta) = k_{\min}(x)$, the stability of an equilibrium is equivalent to the belonging of the point x to the set of attainable equilibria. For rigorous proof, it suffices to show that exactly under $\delta(\alpha, \beta) = \delta_{\max}(x)$ and $k(\alpha, \beta) = k_{\min}(x)$ the function $F_{\alpha,\beta}(y)$ is maximized at any point of the segment $[0, x]$ over all α, β bringing to a stable equilibrium at the point x.

Suppose that, for certain values $\delta_1 \neq \delta_{\max}(x)$ and $k_1 \neq k_{\max}(x)$ implementable under some α and β, we have $\delta_1 + k_1 F(x) = x$ and $k_1 F'(x) < 1$. Our intention is to argue that, whenever $y \leq x$,

$$\delta_{\max} + k_{\min} F(y) \geq \delta_1 + k_1 F(y).$$

Rewrite this inequality as $\delta_{\max} - \delta_1 + (k_{\min} - k_1)F(y) \geq 0$.

Using the equilibrium condition $\delta_1 + k_1 F(x) = \delta_{\max} + k_{\min} F(x) = x$, express $(k_{\min} - k_1)$ and substitute the result into the inequality to obtain

$$(\delta_{\max} - \delta_1)\left(1 - \frac{F(y)}{F(x)}\right) \geq 0.$$

In the case $y \leq x$, the desired inequality follows from implementability of $\delta_1 < \delta_{\max}$ and monotonicity of a distribution function.•

Example 5.5 The paper [13] and Chap. 3 explored real social networks *Facebook*, *Livejournal* and *Twitter*, demonstrating that $F(x)$ can be approximated by a function from the family

$$F(x, \theta, \lambda) = \frac{\text{arctg}(\lambda(x - \theta)) + \text{arctg}(\lambda\theta)}{\text{arctg}(\lambda(1 - \theta)) + \text{arctg}(\lambda\theta)}. \tag{5.17}$$

Here the parameter θ characterizes a phenomenon occurring in a network and causing conformity behavior with the binary actions; this parameter is independent

Fig. 5.5 The values of
$x^*(\alpha, \beta)$ in Model I for social
network Facebook ($\theta = 0.5$,
$\lambda_F = 13$) under the
distribution function (5.17)

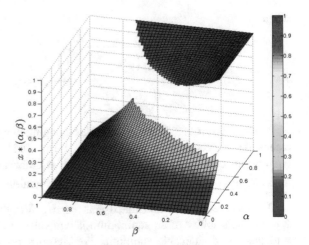

of the network structure. On the other part, the parameter λ characterizes the
connection graph of a social network, being independent of the above phenomenon.
The best approximation of the parameter λ: $\lambda_F \approx 13$ has been found above.
Figure 5.5 illustrates the relationship between the CBE and the actions of the two
Principals.

According to Fig. 5.5, the attainability set is not the segment [0, 1]. In other
words, far from any state of *Facebook* can be implemented as the CBE in the game
of the two Principals.

Introduce the notation $V_{\alpha,\beta}^{\lambda} = \{(x, \theta) : x \in V_{\alpha,\beta} \text{ for } F(x, \theta, \lambda)\}$, $W_{\alpha,\beta}^{\lambda} = \{(x, \theta) :$
$x \in W_{\alpha,\beta} \text{ for } F(x, \theta, \lambda)\}$, where $F(x, \theta, \lambda)$ is defined by (5.17). For different values
of λ, Figs. 5.6 and 5.7 show these sets constructed during the numerical
experiments.

According to the definition of $W_{\alpha,\beta}^{\lambda}$, the attainability set of the social network
described by (5.17) represents a section of the set $W_{\alpha,\beta}^{13}$ under fixed θ. From Fig. 5.7
corresponding to *Facebook* with $\lambda = 13$ and $\theta = 0.5$, we find $W_{\alpha,\beta} \approx [0, 0.4) \cup$
(0.8, 1]. This result agrees with Fig. 5.5.

Now, consider informational confrontation within the framework of Model II,
i.e., add k provokers and l immunizers (the sets K and L, respectively) to the set N of
conformists. Then the probability that an arbitrarily chosen agent has a threshold
not exceeding $x < 1$ comprises the probabilities of two independent events, namely,

(1) the probability that the chosen agent is a provoker $\left(\frac{k}{k+l+n}\right)$;

(2) the probability that the chosen agent is a conformist and his threshold does not
exceed $x < 1$ $\left(\frac{n}{k+l+n} F(x)\right)$.

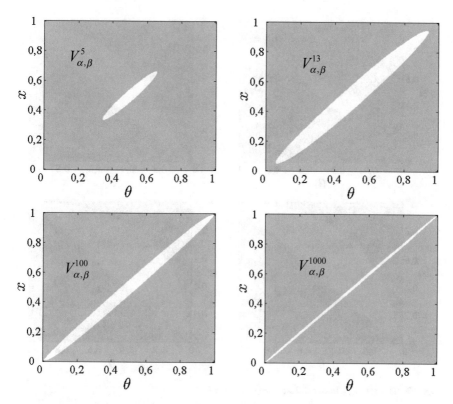

Fig. 5.6 The set $V_{\alpha,\beta}^{\lambda}$ under different λ (*in grey color*)

Introduce the following notation for the proportions of provokers and immunizers in the total number of the agents:

$$\alpha' = \frac{k}{k+l+n}; \beta' = \frac{l}{k+l+n}. \qquad (5.18)$$

Obviously, $\alpha' + \beta' < 1$. In some sense, the quantities α' and β' match the probabilities α and β in Model I of the informational confrontation between the two Principals (see the beginning of this subsection). The probability α that an arbitrarily chosen agent is excited (in Model I) equals the probability that an arbitrarily chosen agent appears an external "provoker" (in Model II). The similar analogy exists between the probabilities β and β'.

Recall that the threshold of an arbitrarily chosen agent is surely smaller than 1. Therefore, we have obtained the new set $N \cup K \cup L$ of the agents whose thresholds

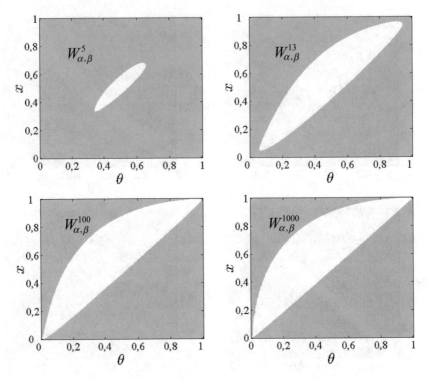

Fig. 5.7 The set $W_{\alpha,\beta}^{\lambda}$ under different λ (*in grey color*)

form independent identically distributed variables with the following distribution function matching Model II:

$$F_{KL}(\alpha', \beta', x) = \begin{cases} \alpha' + (1 - \alpha' - \beta')F(x), & 0 \leq x < 1, \\ 1, & x = 1. \end{cases} \tag{5.19}$$

The distribution functions (5.10) and (5.19) differ; accordingly, Models I and II of the present subsection are not equivalent.

Assertion 5.8 In Model II, the attainability set is $W_{KL} = (0, 1]$. If $F(0) = 0$, then $W_{KL} = [0, 1]$.

Proof of Assertion 5.8. The point 1 is attained due to (5.19).

Let $x_1 \in (0, 1)$. In the case $x_1 \in \{x : F(x) = 0\}$, for the equality $F_{KL}(\alpha'_1, \beta'_1, x_1) = x_1$ to hold true it is necessary and sufficient that $\alpha'_1 = x_1$ and β'_1 is any value satisfying $\alpha'_1 + \beta'_1 < 1$. The left-hand derivative of the function $F_{KL}(\alpha'_1, \beta'_1, x)$ exists owing to monotonicity of the distribution function, taking the value $F'_{KL+}(x_1) = 0$ at the point x_1. In other words, the curve $F_{KL}(\alpha'_1, \beta'_1, x)$ crosses the bisecting line "from left and top." If $\{x : F(x) = 0\} = \emptyset$, then the point $x_1 = 0$ is unattainable, since $\alpha' + \beta' < 1$.

Fig. 5.8 The equilibria in
Model II of social network
Facebook ($\theta = 0.5$, $\lambda_F = 13$)
under the distribution function
(5.17): k provokers and
l immunizers added to
$n = 100$ agents

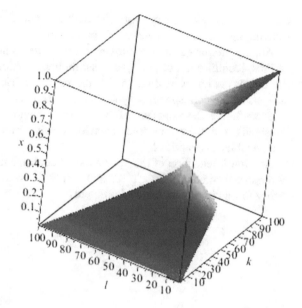

Under $F(x_1) > 0$, it is possible to choose a small value $\varepsilon > 0$ so that $x_1 - \varepsilon \geq 0$,
$1 - x_1 + \varepsilon \left(1 - \frac{1}{F(x_1)}\right) \geq 0$, $1 - \frac{\varepsilon}{F(x_1)} > 0$ and $\varepsilon F'_+ (x_1) < F(x_1)$.

Set $\alpha'_1 = x_1 - \varepsilon \geq 0$, see the first inequality above, and solve the equation
$F_{KL}(\alpha'_1, \beta', x_1) = x_1$ for β'. This gives $\beta' = \beta'_1 = 1 - x_1 + \varepsilon \left(1 - \frac{1}{F(x_1)}\right)$.

The above system of inequalities implies $\beta' \geq 0$ and $\alpha' + \beta' < 1$. Therefore, we
have constructed the distribution function $F_{KL}(\alpha'_1, \beta'_1, x)$ that crosses the bisecting
line at the point x_1 "from left and top."•

Figure 5.8 demonstrates the equilibria in Model II of social network *Facebook*
($\theta = 0.5$, $\lambda_F = 13$) under the distribution function (5.17).

The analysis results of the simultaneous "opposite" impacts on a network
structure, viz., equilibria monotonicity in the control actions (Assertions 5.3 and
5.4) and the structural properties of the attainability sets (Assertions 5.5–5.7) will be
used for the game-theoretic study of informational confrontation models in
Chap. 10.

Thus, we have suggested a *macrodescription* (in terms of Chap. 3) for the
threshold behavior of a mob affected by control actions. An essential advantage of
such stochastic models is a simple analytical form of the relationship between the
distribution functions of the agents' thresholds (*ergo*, the equilibrium mob states)
and the control actions applied. This "simplicity" allows posing and solving control
problems (mob "excitation" and "immunization"), as well as analyzing the informational confrontation between the subjects performing mob control. Two types of
control have been considered, namely, control of the agents' thresholds and staff

control. In the elementary case, the derived control models are equivalent, yet differing appreciably in the case of confrontation.

Among promising lines of further research, we mention the following.

First, identification of the typical distribution functions of the agents' thresholds (by analogy with expression (5.17), the results in Example 5.5 and so on) and analysis of the corresponding typical control actions.

Second, the development of collective behavior models where the agents' thresholds and their threshold behavior involve more general (yet, practically interpretable) assumptions.

And third, reduction of the informational confrontation problem in mob control to a game-theoretic setting, in order to apply the whole arsenal of modern game theory to this important class of problems.

Chapter 6
Game-Theoretic Models of Mob Control

This paper studies models of centralized, decentralized and distributed control of excitation in a network of interacting purposeful agents [73].

Problem description. Consider a set of interconnected agents having mutual influence on their decision-making. Variations in the states of some agents at an initial step accordingly modify the state of other agents. The nature and character of such dynamics depend on the practical interpretation of a corresponding network. Among possible interpretations, we mention the propagation of excitation in biological networks (e.g., neural networks) or in economic networks [49, 54], failure models (in the general case, structural dynamics models) in information and control systems and complex engineering systems, models of innovation diffusions, information security models, penetration/infection models, consensus models and others, see an overview in [48].

The control problem of the purposeful "excitation" of a network is to find a set of agents for applying an initial control action so that the network reaches a required state. This abstract statement covers informational control in social networks [8, 48], collective threshold behavior control, etc.

Network. Let $N = \{1, 2, \ldots, n\}$ be a finite set of *agents*; they form an ANS described by a directed graph $\Gamma = (N, E)$, where $E \subseteq N \times N$ denotes the set of arcs. Each agent is in one of two states, "0" or "1" (passivity or activity, being unexcited or excited, respectively). Designate by $x_i \in \{0; 1\}$ the state of agent i ($i \in N$) and by $\bar{x} = (x_1, x_2, \ldots, x_n)$ the vector of the agents' states. For convenience, the transition from passivity to activity is called the "excitation" of an agent.

Behavior of agents. Assume that, initially, all agents appear passive and the network dynamics is described by a mapping $\Phi: 2^N \to 2^N$. Here $\Phi(S) \subseteq N$ indicates the set of agents having state "1" at the end of the transient process caused by network "excitation." In this section, such excitation represents a variation (switching from passivity to activity) in the states of agents from a set (*coalition*) $S \subseteq N$ that takes place at the initial step. By assumption, the control actions are applied one time.

© Springer International Publishing AG 2017
V.V. Breer et al., *Mob Control: Models of Threshold Collective Behavior*,
Studies in Systems, Decision and Control 85,
DOI 10.1007/978-3-319-51865-7_6

Concerning the mapping $\Phi(\cdot)$, suppose that it enjoys the following properties:

A.1 (reflexivity). $\forall\, S \subseteq N$: $S \subseteq \Phi(S)$;

A.2 (monotonicity). $\forall\, S, U \subseteq N$ such that $S \subseteq U$: $\Phi(S) \subseteq \Phi(U)$.

A.3 (convexity). $\forall\, S, U \subseteq N$ such that $S \cap U = \varnothing$: $\Phi(S) \cup \Phi(U) \subseteq \Phi(S \cup U)$.

For a given mapping $\Phi(\cdot)$, it is possible to define a function $\hat{G}: \{0; 1\}^n \to \{0; 1\}^n$ that associates the vector $\bar{x}$ of the initial states of the agents with the vector of their final states:

$$\hat{G}_i(\bar{x}) = \begin{cases} 1, & \text{if } i \in \Phi(\{j \in N | x_j = 1\}) \\ 0, & \text{otherwise.} \end{cases}$$

Similarly, we easily define the states of agents "excited indirectly":

$$G_i(\bar{x}) = \begin{cases} 1, & \text{if } \hat{G}_i(\bar{x}) = 1 \text{ and } x_i = 0 \\ 0, & \text{otherwise,} \end{cases}, i \in N.$$

The section is organized as follows. Section 6.1 poses the centralized control problem of network excitation, focusing on a special case with the threshold behavior of the agents. Section 6.2 gives a formal statement of the decentralized control problem where the agents make independent decisions on their "excitation." Moreover, the implementability of the efficient or given states of a network is investigated from the game-theoretic view. The problem of mob control serves as a possible application. And finally, we study a model of the informational confrontation of subjects that apply control actions to the network, being interested in its certain states.

6.1 Centralized Control Problem

Consider given functions $C: 2^N \to \mathfrak{R}^1$ and $H: 2^N \to \mathfrak{R}^1$. The former characterizes the costs $C(S)$ of the initial variation in the states of agents from a coalition $S \in 2^N$, while the latter describes the income $H(W)$ from the resulting "excitation" of a coalition $W \in 2^N$. The subjects incurring these costs depend on a specific statement of the problem.

The goal function of a control subject (*Principal*) is the difference between the income and the costs. For the Principal, the *centralized control problem* lies in choosing a set of initially excited agents to maximize the goal function $v(S)$:

$$v(S) = H(\Phi(S)) - C(S) \to \max_{S \subseteq N}, \tag{6.1}$$

In this setting, the Principal incurs the costs and receives the income.

In the general case (without additional assumptions on the properties of the functions $C(\cdot)$ and $H(\cdot)$, and on the mapping $\Phi(\cdot)$), obtaining a solution $S^* \subseteq N$ of the discrete problem (6.1) requires exhausting all 2^n possible coalitions. The design of efficient solution methods for this problem makes an independent field of investigations (we refer to [42, 74], etc. for several successful statements of optimization problems for system's staff employing rather simple algorithms). The ANS state S^* maximizing the goal function (6.1) will be called *efficient*.

A special case engages the cost function and income function being additive with respect to the agents:

$$u(S) = \sum_{i \in \Phi(S)} H_i - \sum_{j \in S} c_j, \tag{6.2}$$

where $(c_i, H_i)_{i \in N}$ are known nonnegative constants. For the time being, we deal with the additive case (6.2) for simplicity.

In the centralized control problem, the agents (network nodes) are passive in some sense. Notably, the Principal "excites" the agents from a set S, and then this excitation propagates according to the operator $\Phi(\cdot)$.

Alternative formulations of the control problem are possible, e.g., income maximization under limited costs (the so-called knapsack problem if the cost function and income function enjoy the additive property with respect to the agents) or costs minimization for a given income.

The discrete problems of form (6.1) suffer from a high computational complexity for the large ANSs (i.e., the networks with very many agents). Therefore, in such cases, the networks are treated as random graphs with specified probabilistic characteristics and the control problem is stated in terms of expected values (e.g., optimization of the expected number of excited agents).

An example is threshold behavior. Imagine agents in an ANS having mutual *influence* on each other. Arc (i, j) from node i to node j corresponds to the influence of agent i on agent j (no loops are admissible). Denote by $N^{in}(i) = \{j \in N \mid \exists (j; i) \in E\}$ the set of "*neighbors*," i.e., the agents having an influence on agent i ("initiators"). By analogy, let $N^{out}(i) = \{j \in N \mid \exists (i; j) \in E\}$ designate the set of agents ("followers") being influenced by agent i, with $n^{out}(i) = |N^{out}(i)|$ and $n^{in}(i) = |N^{in}(i)|$.

The process of collective decision-making by the agents can be described through different models (see the surveys in [48, 56, 86], de Groot's consensus model [33]). Consider a special case of threshold behavior adopted by the agents:

$$x_i^t = \begin{cases} 1, & \frac{1}{n-1} \sum\limits_{j \in N^{in}(i)} x_j^{t-1} \geq \theta_i \\ x_i^{t-1}, & \text{otherwise} \end{cases}. \tag{6.3}$$

Here x_i^t is the state of agent i at step t, and $\theta_i \in [0, 1]$ represents the threshold of this agent, $t = 1, 2, \ldots$ The initial conditions are given: $x_i^0 = x_i, i \in N$. Model (6.3) presupposes that, being once excited (by a control action or under the impact of excited neighbors), an agent never becomes "passive." Obviously, any graph Γ with dynamics (6.3) enjoys the following properties. The number of active agents forms a nondecreasing function of time, the transient process terminates at most after n steps, the correspondence between the initial and final states meets A.1–A.3.

It seems interesting to study a special case of the *threshold behavior with unity thresholds* ($\theta_i = 1$, $i \in N$). In other words, an agent becomes excited from, at least, one of his initiators. The corresponding mapping $\Phi(\cdot)$ is *"linear"*: $\forall S, U \subseteq N$ such that $S \cap U = \varnothing$ we have $\Phi(S) \cup \Phi(U) = \Phi(S \cup U)$, i.e., $\Phi_0(S) = \bigcup_{i \in S} \Phi_0(\{i\})$.

Function (6.2) appears superadditive, viz., $\forall S, U \subseteq N$ such that $S \cap U = \varnothing$ we obtain $u(S \cup U) \leq u(S) + u(U)$.

Let the graph Γ be acyclic and the specific costs of the initial excitation of any agent make up c. For each agent $i \in N$, find the set M_i of all his *"indirect followers,"* i.e., the agents (including agent i) connected to this agent via paths in the graph. Evaluate $h_i = \sum_{j \in M_i} H_j$, which is the payoff from the excitation of agent i. Owing to the acyclic property of the graph and the homogeneous costs, a reasonable strategy lies in the excitation of the agents having no initiators (denote the set of such agents by $M \subseteq N$). Then problem (6.1) acquires the form $\sum_{j \in \bigcup_{i \in S} M_i} H_j - c\,|S| \to \max_{S \subseteq N}$.

Despite a series of simplifying assumptions (threshold behavior, unit thresholds, homogeneous costs and acyclic graph), the resulting centralized control problem still admits no analytical solution and requires exhausting all subsets of the set M. It is possible to apply different heuristics, e.g., believing that the "optimal" excitation covers the agents whose payoff exceeds the costs: $S^* = \{i \in M \mid h_i \geq c\}$. Another approach is to sort the agents from the set M in the descending order of h_i, adding them to the desired set starting from agent 1 (until the corresponding increase in the "payoff" becomes smaller than the specific costs).

Therefore, the centralized control problems of network excitation have simple solutions merely in some cases. Now, we concentrate on possible statements and solution methods of the *decentralized control problems* where the agents make independent decisions on their "excitation."

6.2 Decentralized Control Problems

Consider a given *mechanism* (decision-making procedure [59, 74]) $\sigma = \{\sigma_i(G(\bar{x})) \geq 0\}_{i \in N}$ of payoff allocation among the agents. According to the mechanism σ, additional payoffs can be received only by the agents excited at the initial step. Allocation concerns the "payoff" from the indirect excitation of other agents.

Suppose that the agents represent active subjects in the following sense. Under the known mechanism σ, at the initial step they simultaneously and independently choose their states (activity or passivity). Further dynamics of the agents' states is still described by the operator $\Phi(\cdot)$.

The goal function of agent i (denoted by $f_i(\bar{x})$) makes the difference between his payoff and costs:

$$f_i(\bar{x}) = \sigma_i(G(\bar{x})) + (H_i - c_i)x_i, \ i \in N. \tag{6.4}$$

If an agent decides to be excited, he incurs the excitation costs and obtains the payoff H_i plus an additional incentive from the Principal (according to the payoff allocation mechanism). *Self-excitation* is the profitability of agent's excitation regardless of the payoff allocation mechanism. It follows from (6.4) that, if $H_i > c_i$, then agent i self-excites. In the sequel, we believe that $c_i > H_i$, $i \in N$, to avoid the "self-excitation" of the active agents (except other conditions are clearly stated).

There exist different payoff allocation mechanisms for the agents. Typical examples are *balanced* mechanisms (the total agents' payoff equals the Principal's payoff from "indirect excitation"), namely, the *uniform allocation mechanism*

$$\sigma_i(G(\bar{x})) = \frac{\sum\limits_{j \in N} H_j G_j(\bar{x})}{\sum\limits_{j \in N} x_j} x_i, \ i \in N, \tag{6.5}$$

the mechanism of costs-proportional allocation

$$\sigma_i(G(\bar{x})) = \frac{\sum\limits_{j \in N} H_j G_j(\bar{x})}{\sum\limits_{k \in N} c_k x_k} c_i x_i, \ i \in N, \tag{6.6}$$

and *the mechanism of limit contribution-proportional allocation*

$$\sigma_i(G(\bar{x})) = \frac{[\sum\limits_{j \in N} H_j G_j(\bar{x}) - \sum\limits_{j \in N} H_j G_j(\bar{x}_{-i}, 0)]}{\sum\limits_{k \in N} x_k [\sum\limits_{j \in N} H_j G_j(\bar{x}) - \sum\limits_{j \in N} H_j G_j(\bar{x}_{-k}, 0)]} \sum\limits_{j \in N} H_j G_j(\bar{x}) x_i, \ i \in N. \tag{6.7}$$

According to expressions (6.4)–(6.7), the Principal reallocates the payoff from the indirect excitation of other agents by the initially excited agents among the latter. Moreover, indirect excitation causes no costs.

Equilibrium and efficient states of network. By definition, a vector $\bar{x}^*$ is a *Nash equilibrium* [64] in the agents' game if

$$\begin{aligned} \sigma_i(G(\bar{x}^*)) + (H_i - c_i)x_i^* \\ \geq \sigma_i(G(\bar{x}_{-i}^*, 1 - x_i^*)) + (H_i - c_i)(1 - x_i^*), \ i \in N. \end{aligned} \tag{6.8}$$

In a Nash equilibrium, the active agents benefit nothing by switching their state to passivity (provided that the other agents keep their states unchanged), whereas the passive agents benefit nothing by switching their states to activity. In other words, the active agents (i.e., the agents $i \in N$ such that $x_i^* = 1$) obey the inequality

$$\sigma_i(G(\bar{x}^*)) + H_i \geq c_i, \tag{6.9}$$

On the other hand, the passive agents (i.e., the agents $i \in N$ such that $y_i^* = 0$) obey the inequality

$$\sigma_i(Gx_{-i}^*, 1) + H_i \leq c_i. \tag{6.10}$$

Write down and analyze conditions (6.9)–(6.10) for mechanism (6.5). Under the uniform allocation mechanism, in the equilibrium the active agents and the passive agents satisfy the conditions

$$\frac{\sum\limits_{j \in N} H_j G_j(\bar{x}^*)}{\sum\limits_{j \in N} x_j^*} + H_i \geq c_i, \tag{6.11}$$

and

$$c_i \geq \frac{\sum\limits_{j \in N} H_j G_j(x_{-i}^*, 1)}{\sum\limits_{j \in N} x_j^* + 1} + H_i, \tag{6.12}$$

respectively.

What is the connection between the solution set of the centralized control problem and the set of Nash equilibria? When is it possible to *implement* the efficient state of the network (see criteria (6.1) and (6.2)) by centralized control? Which hypotheses are true: (a) an efficient state forms a Nash equilibrium, (b) at least, one equilibrium state appears efficient? Consider two simple examples showing that the sets of efficient states and Nash equilibria have nontrivial connection and both hypotheses generally fail.

Example 6.1 There are three agents with unit thresholds (see (6.3)). Their costs and payoffs are presented below.

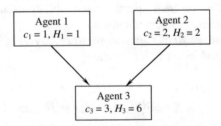

The list of the efficient state vectors comprises $(0; 1; 0)$, $(1; 0; 0)$, and $(1; 1; 0)$. Regardless of the payoff allocation mechanism, the Nash equilibria are the rest nontrivial state vectors. Therefore, all four Nash equilibria appear inefficient and all efficient states are nonequilibrium states. •

Example 6.2 Within the framework of Example 6.1, we modify the costs and payoff of agent 3 as follows:

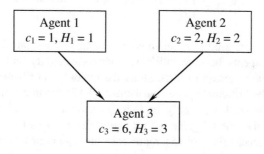

Again, the efficient state vectors are (0; 1; 0), (1; 0; 0), and (1; 1; 0). The Nash equilibrium–the vector (1; 1; 0)–is unique and efficient. •

And so, the choice of the payoff allocation mechanisms for the decentralized implementation of efficient network states remains unsettled in the general case.

Mechanisms (6.5)–(6.7) are balanced, i.e., the Principal allocates the whole (!) payoff from "indirect network excitation" among the initially active agents. Such mechanisms can be called *motivational control mechanisms* (see [74])–the Principal stimulates the agents to choose certain states as the equilibria of their game. An alternative approach lies in stronger *institutional control* [74], where the Principal establishes certain constraints and norms of the agents' activity. We examine a corresponding model in the network excitation problem.

Institutional control. Let the goal functions of the agents have the form

$$f_i(\bar{x}) = s_i(\bar{x}) + (H_i - c_i)x_i, \tag{6.13}$$

where $s_i(\bar{x}) \geq 0$ is an incentive paid by the Principal to agent i ($i \in N$). Generally, this incentive depends on the state vector (action profile) of all agents. Following [59, 74], the structure $s(\bar{x}) = \{s_i(\bar{x})\}_{i \in N}$ will be termed the vector-form *incentive function of the agents* from the Principal. The dependence of the agents' incentives on the action profile of all agents is a special case of the above payoff allocation mechanism.

Fix some set V of agents. Consider the following "institutional control" problem. Find the minimal vector-form incentive function (in the sense of the Principal's total costs) that "excites" the given set V of agents as a Nash equilibrium $\bar{x}^*(s(\cdot))$ in their game. Formally, this problem admits the following statement:

$$\begin{cases} \sum_{i \in N} s_i(x^*) \to \min_{s(\cdot)} \\ x_i^*(s(\cdot)) = 1, \ i \in V, \\ x_j^*(s(\cdot)) = 0, \ j \notin V. \end{cases} \tag{6.14}$$

At the first glance, problem (6.14) seems complicated; however, it has a rather easy solution using the *decomposition theorem of the agents' game*. Consider the following vector-form incentive function:

$$s_i^*(\bar{x}) = \begin{cases} c_i - H_i + \varepsilon_i, & \text{if } i \in V \text{ and } x_i = 1 \\ 0, & \text{otherwise} \end{cases}, \qquad (6.15)$$

where ε_i ($i \in V$) are arbitrarily small strictly positive constants. If we nullify these constants, the agents become indifferent between activity and passivity, and all conclusions below remain in force under the hypothesis of benevolence [74].

Clearly, if the Principal applies mechanism (6.15), the unit actions chosen by the agents from the set V (and only by them!) forms a unique *dominant strategies' equilibrium* in the agents' game with the goal functions (6.13).

Moreover, mechanism (6.15) represents the ε_V-optimal solution of problem (6.14), where $\varepsilon_V = \sum_{i \in V} \varepsilon_i$.

We make an important terminological remark. Formally, problem (6.14) is a motivational control problem; nevertheless, its solution (6.15) can be interpreted as institutional control. The Principal establishes rather strict norms of activity for the agents: any deviation from the assigned behavior causes penalties (their incentives vanish).

The Principal's goal function $F(\cdot)$ is the difference between the payoff from exciting the set $\Phi(V)$ of agents and the total costs (6.15) to implement the excitation of the agents from the set V (see formula (6.2)). In other words,

$$F(V) = \sum_{i \in \Phi(V)} H_i - \sum_{j \in V} s_j(y^*) = \sum_{i \in \Phi(V)} H_i + \sum_{j \in V} H_j - \sum_{i \in V} c_i - \varepsilon_V. \qquad (6.16)$$

The comparison of expressions (6.16) and (6.2) brings to the following conclusion. Under sufficiently small values of ε_V (see the discussion above), one obtains the condition $F(V) \geq u(V)$. It would seem that the decentralized control problem is completely solved! Really, we have constructed a partial solution of this problem, since in mechanisms (6.5)–(6.7) the Principal does not explicitly specify the actions expected from the agents. Thus, the agents can play their game "independently" (the basic idea of control decentralization consists in designing a certain procedure of the autonomous interaction of agents that leads to the choice of the most efficient action vector in the sense of some centralized criterion). "Mechanism" (6.15) explicitly states the actions expected by the Principal from different agents. Furthermore, the Principal still has to solve the centralized control problem (6.1). Notably, knowing his optimal payoff (6.16) (in the sense of the minimal costs to motivate the excited agents), the Principal has to define a coalition for initial excitation:

$$F(V) \to \max_{V \subseteq N}. \tag{6.17}$$

Therefore, a positive side of the decentralized mechanisms is that the agents may possess incomplete information (they do not have to compute the Nash equilibrium (6.8) or solve the discrete optimization problems (6.1) or (6.17)). And a drawback is the complexity (or even infeasibility) of constructing an efficient decentralized mechanism.

Among the advantages of the centralized mechanism, we mention that all "cognitive" (informational and computational) costs belong to the Principal; however, these costs may appear appreciably high.

As an example, let us consider the mob control problem.

An example: mob "excitation." In the previous sections, we have explored a model of mob control where the agents make decisions (choose between their activity or passivity) depending on the number of active agents. The Principal's efficiency criterion has been the number (or proportion) of active agents. In terms of the model analyzed below, the problem of mob control admits the following statement: choose a set of the initially excited agents to maximize (to minimize, etc.) the number of the indirectly excited agents so that the costs of control satisfy a given budget constraint C_0:

$$\begin{cases} |\Phi(S)| \to \max_{S \subseteq N}, \\ C(S) \le C_0. \end{cases} \tag{6.18}$$

Let the behavior of the agents be described by expression (6.3); moreover, assume that the communication graph of the agents is complete. Renumber the agents in the ascending order of their thresholds: $\theta_1 \le \dots \le \theta_n$. Denote by $P(x) = \frac{1}{n}|\{i \in N : \theta_i < x\}|$ the distribution function of the agents' thresholds and by $\{x_t\}_{t \ge 0}$ the sequence of the proportions of active agents (in discrete time, where t indicates current step).

Suppose that we know the proportion x_0 of the agents acting at step 0. The proportion of the agents whose thresholds do not exceed x_0 makes up $P(x_0)$. And so, at step 1 we have $x_1 = \max \{x_0; P(x_0)\}$. At the next step, the proportion x_2 of active agents is defined by $x_2 = \max \{x_1; P(x_1)\}$ (the thresholds of these agents are not greater than x_1). Arguing by analogy, one easily obtains the following recurrent formula for the behavioral dynamics of the agents (see (2.7)):

$$x_{k+1} = \max \{x_k; P(x_k)\}. \tag{6.19}$$

The equilibria in system (6.19) are determined by the initial point x_0 and the intersection points of the curve $P(\cdot)$ with the bisecting line of quadrant I: $P(x) = x$. The potential stable equilibria are the points where the curve $P(\cdot)$ crosses the bisecting line, approaching it "from above". Figure 6.1 shows an example of the

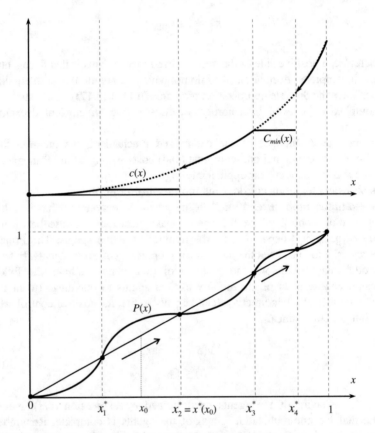

Fig. 6.1 The distribution function of agents' thresholds and cost functions

distribution function of the agents' thresholds in the continuous-time case (see the lower fragment of the figure). The points x_2^* and x_4^* are stable.

Designate by $\Psi(P(\cdot)) \subseteq [0,1]$ the set of roots of the equation $P(x) = x$ (nonempty, since one of the roots is 1). Next, define $x^*(x_0) = \min \{y \in \Psi(P(\cdot)): y > x_0\}$. For instance, the point x_0 in Fig. 6.1 satisfies the condition $x^*(x_0) = x_2^*$. It follows from (6.19) and the equilibrium stability conditions that

$$\Phi(x_0) = \begin{cases} x_0, & \text{if } P(x_0) \leq x_0, \\ x^*(x_0), & \text{if } P(x_0) > x_0. \end{cases} \tag{6.20}$$

Let a nondecreasing function $c(x_0)$ specify the initial excitation costs of a given proportion $x_0 \in [0, 1]$ of the agents (within the framework of our collective behavior model, excitation first applies to the agents with smaller thresholds). By virtue of (6.20), problem (6.18) acquires the form

$$\begin{cases} \Phi(x_0) \longrightarrow \max_{x_0 \in [0,1]}, \\ c(x_0) \leq C_0. \end{cases} \qquad (6.21)$$

Set $x^+ = \max \{x \in [0, 1]: c(x) \leq C_0\}$. Owing to the nondecreasing property of function (6.20), the solution procedure of problem (6.21) seems obvious: the Principal should implement the maximum admissible action x^+ under the existing budget constraints. If, in addition, costs minimization is required, then the optimal solution x_0^* is

$$x_0^* = \begin{cases} x^+, \text{ if } P(x^+) \leq x^+, \\ \max \{y \in \Psi(P(\cdot)): y < x^+\}, \text{if } P(x^+) > x^+. \end{cases} \qquad (6.22)$$

We can solve the inverse problem, i.e., find the minimal costs $C_{\min}(x)$ of an initial excitation such that the final proportion of excited agents in the mob is not smaller than a given level $x \in [0, 1]$. See Fig. 6.1 for an example of its solution.

By assumption, the Principal seeks to maximize the number of excited agents. If the goal lies in mob activity minimization, the corresponding problems are posed and solved by analogy, since the analysis of the stables states and their dependence on the model parameters (see Fig. 6.1) allows characterizing the relationship between the resulting and initial states.

Informational confrontation. Recall the informational confrontation conditions adopted in Sect. 5.3.

Let us state a game-theoretic model of informational confrontation in network excitation control. There are Principals 1 and 2 exerting control actions at the initial step one-time, simultaneously and independently from each other, modifying the initial states of the agents from sets $S_1 \subseteq N$ and $S_2 \subseteq N$, respectively. Assume that we know the relationship $\widehat{\Phi}(S_1, S_2)$ between the final network state and these control actions. The goal functions of the Principals represent the differences between their payoffs and costs: $v_i(S_1, S_2) = H_i(\widehat{\Phi}(S_1, S_2)) - C_i(S_i)$, $i = 1, 2$.

This statement yields a normal-form game of the Principals. If one of them has the right of first move, we obtain a Stackelberg game (or game Γ_1) which can be interpreted as "attack-defense" or "informational influence-counteraction." Possible formulations and solutions of such problems in mob control are postponed to Sect. 10.

Therefore, we have considered the general statement of the network "excitation" problem, establishing the nontrivial character of the decentralized implementation of an efficient equilibrium and suggesting an efficient mechanism of institutional control.

It is possible to identify several directions of promising research. First, the matter concerns analyzing the influence of certain properties of the communication graph of the agents and their decision-making principles on the properties of optimal solutions in control problems. In this section, both these factors (the structure of agents' communication and the models of their cooperative behavior) are "embedded" in the operator $\Phi(\cdot)$. Explicit formulations and exploration of special cases (acyclic graphs, concrete threshold models and other models of agents' decision-making) may yield nontrivial results with practical interpretations.

Second, as far as the standard triad of control types [74] comprises institutional control, motivational control and informational control (the first and second types have been mentioned above), certain interest belongs to analysis of informational control models in network excitation problems.

Third, the presence of a network of agents making strategic decisions suggests involving some results from the theory of cooperative games on graphs (a communication graph restricts the capabilities of coalition formation and interaction among the agents [6, 28, 65, 77]). However, in network excitation problems, communication graphs are connected with the agents' ability to influence other agents (i.e., communication graphs define the goal functions of the agents, *ergo* the characteristic function of the corresponding cooperative games), rather than with the feasibility of forming certain coalitions.

Fourth, it is necessary to study in detail the payoff allocation mechanisms of form (6.5)–(6.7), including their analysis within the framework of social choice theory (e.g., see [59, 62]). Here we mean the characterization of their classes possessing given properties (for instance, implementation of the efficient states as an important property).

And finally, fifth, many prospects are associated with further decentralization of the network "excitation" problem. The approaches of algorithmic game theory [4, 85] and distributed optimization [16, 17, 78] prompt the following idea. We should endeavor to find simple procedures of the local behavior of agents, leading to the optimal state within the initial problem of form (6.1) and (6.17), which has very high algorithmic or computational complexity.

Chapter 7
Dynamic Models of Mob Control in Discrete Time

Written jointly with I.N. Barabanov, this section formulates and solves the mob excitation control problem in the discrete-time setting by introducing an appropriate number of "provokers" at each step of control [9].

The earlier paper [25] and Chap. 5 formulated and solved the static problem of mob excitation control by introducing an appropriate number of "provokers" that was chosen one-time and fixed for the whole time horizon.

Below we involve the basic discrete-time model of mob excitation [25] assuming that the number of "provokers" can be varied at each step. Further exposition has the following organization. Section 7.1 states the mob control problem proper, and Sect. 7.2 analyzes the models with constraints on the number of introduced provokers varying in discrete time. Sects. 7.3 and 7.4 consider the continuous models in the proportion of provokers and the probabilistic models of provokers detection, respectively.

7.1 Mob Control Problem

Suppose that the number of agents is large and the agents are homogeneous in the sense of rule (7.1). The thresholds θ_i represent the realizations of a same random variable. Denote by $F(\cdot)$: $[0, 1] \rightarrow [0, 1]$ the *distribution function* of the agents' thresholds, a nondecreasing function defined on the unit segment (the set of admissible thresholds) that is left-continuous and possesses the right-hand limit at each point of its domain of definition.

Assume that we know the proportion x_k of the active agents at step k, where $k = 0, 1, \ldots$ *Further behavior* of the agents obeys the following recurrent expression (see (2.7)):

© Springer International Publishing AG 2017
V.V. Breer et al., *Mob Control: Models of Threshold Collective Behavior*,
Studies in Systems, Decision and Control 85,
DOI 10.1007/978-3-319-51865-7_7

$$x_{l+1} = F(x_l), l = k, k + 1, \ldots \tag{7.1}$$

According to the results of Chap. 5, adding m_0 provokers to the mob yields a new set of agents whose thresholds form independent identically distributed random variables with the distribution function

$$\hat{F}(x, m_0) = \frac{m_0}{m_0 + n} + \left(1 - \frac{m_0}{m_0 + n}\right) F(x). \tag{7.2}$$

Denote by $x^*(m_0)$ the collective behavior equilibrium of the dynamic system (7.1) corresponding to the distribution function (7.2).

Consider a *fixed time horizon*, i.e., the first K steps. Assuming that, generally, the number of provokers m_k at different steps $k = 0, 1, \ldots, K - 1$ may vary, use expressions (7.1) and (7.2) to construct a dynamic controlled system that describes the proportion of active agents:

$$x_k = \frac{m_{k-1}}{m_{k-1} + n} + \left(1 - \frac{m_{k-1}}{m_{k-1} + n}\right) F(x_{k-1}), \quad k = 0, 1, \ldots, x_0 = 0. \tag{7.3}$$

Here the role of control at step k is played by the number of introduced provokers m_{k-1}.

Within the framework of the dynamic system (7.3), one can formulate and solve different control problems, thereby studying *dynamic models of mob excitation control* (see below).

7.2 Models with Constraints on the Total Number of Provokers

Begin with the elementary case where at each step the number of provokers satisfies the constraint

$$0 \leq m_k \leq M, \ k = 0, 1, \ldots, K - 1, \tag{7.4}$$

while the *efficiency criterion* is defined as the proportion of active agents at step K. Then the corresponding *terminal control problem* takes the form

$$\begin{cases} x_K(m) \xrightarrow[m]{} \max, \\ (3), (4). \end{cases} \tag{7.5}$$

Assertion 7.1. The solution of problem (7.5) is given by $m_k = M, k = 0, 1, \ldots, K - 1$.

Proof of Assertion 7.1. Consider the right-hand side of expression (7.3). By the properties of the distribution function $F(\cdot)$, monotonicity of $\frac{u}{u+n}$ in $u \in N$ and the fact that $\frac{u}{u+n} \leq 1$ for any $u, n \in N$, the expression $\frac{u}{u+n} + \left(1 - \frac{u}{u+n}\right) F(v)$ is monotonically increasing in $u \in N$ and $v \in [0,1]$.

Take step $k = K - 1$. The maximum of x_K is achieved under the maximum possible values of x_{K-1} and $m_{K-1} = M$. By analogy, consider sequentially all preceding steps $0 \leq k < K - 1$ to establish that the optimal choice is also $m_l = M$. •

Corollary. *If* $m = (m_0, \dots, m_{K-1})$ *is a nondecreasing sequence, then* $x_k(m)$ *is a nondecreasing sequence, as well.*

Denote by $l(\hat{x}, m) = \min \{k = 1, \dots, K | x_k(m) \geq \hat{x}\}$ the first step when the proportion of active agents achieves a required value $\hat{x}$ (if the set $\{k = 1, \dots, K | x_k(m) \geq \hat{x}\}$ appears empty, just specify $l(\hat{x}, m) = +\infty$). Within the current model, one can pose the following *performance problem*:

$$\begin{cases} l(\hat{x}, m) \to \min_m, \\ (3), \ (4). \end{cases} \tag{7.6}$$

Assertion 7.2. The solution of problem (7.6) *is given by* $m_k = M, \ k = 0, 1, \dots, K - 1$.

Proof of Assertion 7.2. Fix an arbitrary step $k, 0 < k \leq K$, and an arbitrary step $k', 0 \leq k' < k$. It follows from (7.7) and the proof of Assertion 7.1 that.

$$\forall m_{k'} < M : \quad x_k(M, \dots, m_{k'}, \dots, M) \leq x_k(M, \dots, M, \dots, M). \bullet$$

Problems (7.5) or (7.6) have the following practical interpretation. The Principal benefits most from introducing the maximum admissible number of provokers into the mob at the initial step, doing nothing after that (e.g., instead of first decreasing and then again increasing the number of introduced provokers). This structure of the optimal solution can be easily explained, as in models (7.5) and (7.6) the Principal incurs no costs to introduce and/or keep the provokers.

Let us make an obvious observation concerning M, i.e., the maximum admissible number of provokers. If M satisfies the condition $M \geq n\hat{x}/(1 - \hat{x})$, then the performance problem is solved at one step. Otherwise, there exists a passive mob (a distribution function $F(\cdot)$) such that the value $\hat{x}$ is not achievable in a finite number of steps.

This lower estimate of the maximum number of provokers required seems rather rough, yielding large values in case of large $\hat{x}$. In particular, for $\hat{x} \geq 0.5$ we have $M \geq n$, implying that the number of provokers must exceed the total number of agents in the mob. Formally, the constructed model allows an arbitrary (including arbitrary large) number of provokers. At the same time, clearly, a large number of provokers should be introduced into an almost degenerate mob in the sense of excitability (i.e., when the threshold distribution function is close to zero almost everywhere on [0, 1]).

Now, consider possible statements of the control problems taking into account the Principal's costs.

Models with the Principal's control costs. Under a known nondecreasing cost function $c(m)$, a more general than (7.5) problem is the following *integer dynamic programming problem* with M^K alternatives totally:

$$\begin{cases} H(x_K(m)) - c(m) \to \max_m, \\ (3), (4). \end{cases} \qquad (7.7)$$

Given a fixed "price" $\lambda \geq 0$ of keeping one provoker per unit time, the Principal's costs are defined by

$$c(m) = \lambda \sum_{k=0}^{K-1} m_k. \qquad (7.8)$$

A possible modification of problems (7.7)–(7.8) is the one where the Principal achieves a required proportion $\hat{x}$ of active agents by step K (the *cost minimization problem*):

$$\begin{cases} c(m) \to \min_m, \\ x_K(m) \geq \hat{x}, \\ (3), (4), (8). \end{cases} \qquad (7.9)$$

The forthcoming subsection studies these problems in detail.

7.3 Continuous Control

Let M and n be large. Introduce the notation $\delta_k = \frac{m_k}{m_k + n}$, $\lambda_0 = \lambda n$. Then condition (7.4) acquires the form $\delta_k \in [0, \Delta], k = 0, 1, \ldots, K-1$, where $\Delta = \frac{M}{M+n}$, and costs (7.8) can be rewritten as

$$c(\delta) = \lambda_0 \sum_{k=0}^{K-1} \frac{\delta_k}{1 - \delta_k}, \qquad (7.10)$$

where $\delta = (\delta_0, \delta_1, \ldots, \delta_{K-1})$, δ_k specify real numbers, $k = 0, 1, \ldots, K-1$.

According to the accepted notation, the dynamic system (7.3) admits the representation

$$x_k(\delta) = \delta_{k-1} + (1 - \delta_{k-1})F(x_{k-1}), k = 0, 1, \ldots, x_0 = 0. \qquad (7.11)$$

We endeavor to explore the properties of the *attainability set*

$$D = \bigcup_{\delta:\, 0 \leq \delta_k \leq \Delta,\, k=\overline{0,K-1}} x_K(\delta) \subseteq [0;1].$$

By analogy with the proof of Assertion 7.1, it is possible to show that $x_K(\delta)$ enjoys monotonicity in δ_k, $k = 0, 1, \ldots, K-1$. If this relationship is continuous, then (7.11) defines a continuous monotonic mapping $[0, \Delta]^K$ into $[0, 1]$ (monotonicity in the sense of cone). That is, we have the following result.

Assertion 7.3. If the distribution function $F(.)$ is continuous, then $D = [0, x_K(\Delta, \ldots, \Delta)]$.

In the sense of potential applications, a major interest is attracted by the case of *unified solutions* $\hat{\delta} = (\rho, \ldots, \rho)$. Here the Principal chooses a same proportion of provokers $\rho \in [0, \Delta]$ at all steps. Let z $D_0 = \bigcup_{\rho \in [0;\Delta]} x_K(\rho, \ldots, \rho) \subseteq [0; 1]$ be the attainability set in the unified case. Since $x_K(\rho, \ldots, \rho)$ is a monotonic continuous mapping of $[0, \Delta]$ into $[0, 1]$ such that $x_K(0, \ldots, 0) = 0$, then $D_0 = D$ and we get the following assertion.

Assertion 7.4. If the distribution function $F(\cdot)$ is continuous, then for any admissible sequence δ there exists an equivalent unified solution $\hat{\delta}$, i.e., an admissible $\hat{\delta}$ such that $x_K(\delta) = x_K(\hat{\delta})$.

As a matter of fact, Assertion 7.4 means that the Principal may confine itself to the unified solutions subject to the attainability set.

Assertion 7.4 appears nonconstructive, as merely stating the existence of an equivalent unified solution without methods to find it. In a series of special cases (including Example 7.1), it is possible to obtain an explicit expression for the proportion of active agents at step k, as well as the equivalent unified solution.

Example 7.1. Consider the uniform distribution of the agents' thresholds, i.e., $F(x) = x$. Take an arbitrary step l: $0 \leq l \leq K - 1$ and an arbitrary time increment s: $l + s \leq K$. Using inductive proof, one can demonstrate that

$$x_{l+s} = 1 + (x_l - 1) \prod_{j=0}^{s-1} (1 - \delta_{l+j}). \tag{7.12}$$

Choose ρ from the condition

$$(1 - \rho)^s = \prod_{j=0}^{s-1} (1 - \delta_{l+j}), \tag{7.13}$$

i.e., $(1-\rho)$ is the geometric mean of the values $(1 - \delta_{l+j})$. Then the initial state x_l yields the same final state x_{l+s} under the control sequence $\{\delta_l, \ldots, \delta_{l+s-1}\}$ and

$\left\{ \underbrace{\rho, \ldots, \rho}_{s} \right\}$. As step l and the time increment are arbitrary, in this fashion one can replace the whole sequence of the control actions δ with the sequence of identical control actions.

If $F(x) = x$, then (7.12)–(7.13) with $x_0 = 0$ directly give

$$x_k(\delta) = \sum_{i=0}^{k-1} \delta_i - \sum_{i,j=0 i \neq j}^{k-1} \delta_i \delta_j + \sum_{i,j,l=0 i \neq j \neq l}^{k-1} \delta_i \delta_j \delta_l - \sum_{i,j,l,s=0 i \neq j \neq l \neq s}^{k-1} \delta_i \delta_j \delta_l \delta_s + \ldots \bullet$$

$$(7.14)$$

Example 7.2. Let $K = 2$ and $F(x) = x$. It appears from (7.14) that $x_1 = \delta_0, x_2 = \delta_0 + \delta_1 - \delta_0 \delta_1$. Choose $H(x) = x$ and the costs defined by (7.10). In this case, problem (7.7) acquires the form

$$\delta_0 + \delta_1 - \delta_0 \delta_1 - \lambda_0 \left[\frac{\delta_0}{1 - \delta_0} + \frac{\delta_1}{1 - \delta_1} \right] \rightarrow \max_{\delta_0, \delta_1 \leq \Delta} .$$

Its solution is described by

$$\delta_0 = \delta_1 = \begin{cases} 0, & \text{if} \quad 1 - \sqrt[3]{\lambda_0} \leq 0, \\ 1 - \sqrt[3]{\lambda_0}, & \text{if} \quad 0 < 1 - \sqrt[3]{\lambda_0} < \Delta, \\ \Delta, & \text{if} \quad 1 - \sqrt[3]{\lambda_0} \geq \Delta. \end{cases}$$

In other words, the solution depends on the price of keeping one provoker per unit time. For a high price, the introduction of provokers becomes even unprofitable.•

Example 7.3. Just like in Example 7.2, choose $F(x) = x$. Consider the costs minimization problem (7.9) with achieving a given value $\hat{x}$ in K steps:

$$\begin{cases} \frac{\delta_0}{1 - \delta_0} + \cdots + \frac{\delta_{K-1}}{1 - \delta_{K-1}} \rightarrow \min_{0 \leq \delta_0, \ldots, \delta_{K-1} \leq \Delta}, \\ 1 - (1 - \delta_0) \cdots (1 - \delta_{K-1}) \geq \hat{x} \end{cases}$$

There exists a solution of this problem if the set of variables in goal function minimization is nonempty. This imposes the constraints $\Delta \geq 1 - \sqrt[K]{1 - \hat{x}}$ on the problem parameters, illustrating that for goal achievement the Principal needs a sufficient number of provokers at its disposal.

Due to monotonicity of the goal function in each variable, the constraint holds as equality. On the other hand, the problem has symmetry and the minimum is achieved for the same values δ_i. Therefore, the solution is given by $\delta_0 = \ldots = \delta_{K-1} = 1 - \sqrt[K]{1 - \hat{x}}.\bullet$

Problem (7.7)–(7.8) with the continuous variables takes the following simple form:

$$\begin{cases} H(x_K(\delta)) - c(\delta) \to \max\limits_{\delta}, \\ (10),\ (11), \delta_k \in [0; \Delta]. \end{cases} \tag{7.15}$$

Problem (7.15) can be solved numerically under known payoff functions, cost functions and threshold distribution functions. In particular, Assertion 7.3 leads to a two-step method where step 1 is to find

$$C(z) = \min_{\delta:\ \delta_k \in [0;\Delta],\ x_K(\delta)=z} \sum_{k=0}^{K-1} \frac{\delta_k}{1 - \delta_k}, z \in [0, x_K(\Delta, \ldots, \Delta)],$$

and step 2 is to solve the scalar problem

$$z^* = \arg \max_{z \in [0; x_K(\Delta,\ldots,\Delta)]} [H(z) - \lambda_0 C(z)].$$

Another alternative statement of the mob excitation control problem with $H(x) = x$ is

$$\begin{cases} x_K(\delta) + \sum\limits_{k=0}^{K-1} \left[x_k(\delta) - \lambda_0 \frac{\delta_k}{1 - \delta_k} \right] \to \max\limits_{\delta}, \\ (14), \delta_k \in [0; \Delta]. \end{cases} \tag{7.16}$$

The goal function in problem (7.16) structurally differs from its counterpart in the terminal problem (7.15), incorporating the sum of the proportions of active agents at all steps.

Write the Bellman equation for problem (7.16) with costs (7.10) and $H(x) = x$ (which is assumed in further exposition):

$$B(x, w) = w + (1 - w)F(x) - \lambda_0 \frac{w}{1 - w} \to \max_{w \in [0; \Delta]}. \tag{7.17}$$

Denote by

$$w(x) = \max \left\{ \min \left\{ \Delta; 1 - \sqrt{\frac{\lambda_0}{1 - F(x)}} \right\}; 0 \right\} \tag{7.18}$$

the interior maximum in problem (7.17) yielded by the first-order optimality conditions.

Owing to monotonicity of the distribution function, $w(x)$ represents an nonincreasing function.

The calculation of $B(x, w(x)) = 1 + \lambda_0 - 2\sqrt{\lambda_0(1 - \ominus F(x))}$ shows that the solution of problem (7.17) has three possible "modes" (two boundaries of the segment $[0, \Delta]$ or the internal point (7.18)):

$$w^*(x) = \arg \max_{w \in \{0; w(x); \Delta\}} B(x, w). \tag{7.19}$$

The qualitative analysis of solution (7.19) depending on the parameter λ testifies to the following (recall that this parameter describes the price of keeping one provoker per unit time). For very large values of λ, the optimal solution is the zero number of provokers ($w = 0$); for very small λ, the maximum possible number of provokers ($w = \Delta$). If λ takes "intermediate" values, then the optimal solutions belong to the interval $(0, \Delta)$.

Designate by δ^* the solution of problem (7.16). The dynamic system (7.11) has the following property.

Assertion 7.5. If the inequality

$$F(x) \geq 1 - (1 - x)^2 / \lambda_0, \tag{7.20}$$

holds for all $x \in [0, x^*(\Delta)]$, then $x_k(\delta^*)$ is a nondecreasing sequence in k.

Proof of Assertion 7.5. Substituting expression (7.18) into (7.11) gives the monotonicity condition

$$1 - \sqrt{\frac{\lambda_0}{1 - F(x)}} + \sqrt{\frac{\lambda_0}{1 - F(x)}} F(x) \geq x,$$

After trivial transformations, we arrive at (7.20).•

Assertion 7.6. Assume that for all $\alpha, \beta, \gamma \in [0; 1]$: $\alpha \geq \beta, \gamma \leq x^*(\Delta)$, the distribution function satisfies the condition

$$\beta + (1 - \beta)F(\alpha + (1 - \alpha)F(\gamma)) \leq \alpha + (1 - \alpha)F(\beta + (1 - \beta)F(\gamma)). \tag{7.21}$$

Then there exists an optimal nondecreasing sequence δ^*.

Proof of Assertion 7.6. Let $\exists l : 0 < l < K - 1$ and $\delta_l^* > \delta_{l+1}^*$.
If it is possible to find $\delta_l[0, \Delta] \in$ and $\delta_{l+1} \in [0, \Delta]$ such that
(A1) $\delta_l \leq \delta_{l+1}$,
(A2) $x_{l+2}(x_l, \delta_l, \delta_{l+1}) = x_{l+2}(x_l, \delta_l^*, \delta_{l+1}^*)$,
(A3) $c(\delta^*) \geq c\delta_0^*, \ldots, \delta_{l-1}^*, \delta_l, \delta_{l+1}, \delta_{l+2}^*, \ldots, \delta_{K-1}^*)$,
then $\delta_0^*, \ldots, \delta_{l-1}^*, \delta_l, \delta_{l+1}, \delta_{l+2}^*, \ldots, \delta_{K-1}^*$ also gives the solution of problem (7.15).
Choose $\delta_l = \delta_{l+1}^*$, $\delta_{l+1} = \delta_{l+1}^* + \varepsilon$, where $\varepsilon \in [0; \delta_l^* - \delta_{l+1}^*]$. As a result, (A1) holds. Condition (A3) is satisfied due to the monotonicity of the cost function, while expression (A2) yields

$$\delta_{l+1} + (1 - \delta_{l+1})F(\delta_l + (1 - \delta_l)F(x_l)) = \delta^*_{l+1} + (1 - \delta^*_{l+1})F(\delta^*_l +$$
$$(1 - \delta^*_l)F(x_l)), \text{i.e.,}$$

(A4) $\varepsilon = \dfrac{(1-\delta^*_{l+1})[F(\delta^*_l + (1-\delta^*_l)F(x_l)) - F(\delta^*_{l+1} + (1-\delta^*_{l+1})F(x_l))]}{1 - F(\delta^*_{l+1} + (1-\delta^*_{l+1})F(x_l))}$.

Note that, for $F(x) = x$, $\varepsilon \equiv \delta^*_l - \delta^*_{l+1}$.

It follows from (A4) that $\varepsilon \geq 0$, and the condition $\varepsilon \leq \delta^*_l - \delta^*_{l+1}$ can be rewritten as

$$\delta^*_{l+1} + (1 - \delta^*_{l+1})F(\delta^*_l + (1 - \delta^*_l)F(x_l)) \leq \delta^*_l + (1 - \delta^*_l)F(\delta^*_{l+1} +$$
$$(1 - \delta^*_{l+1})F(x_l)).$$

By $x_l \leq x^*(\Delta)$ and condition (7.21), the last inequality holds.

The above reasoning shows that the decreasing segment of the two neighbor elements in the control sequence δ can be replaced with an "equivalent" nondecreasing one, still retaining the values of x and the optimality condition. It suffices to argue that, using this operation a finite number of times, we make the whole sequence δ nondecreasing. Really, let δ^*_m be the minimum value from the whole sequence. Apply the described procedure a finite number of times (at most, K) to move this value to the zero position in the new sequence. Next, find the second minimum value and move it to the first position via at most $(K - 1)$ permutations, and so on. Therefore, the number of permutations is bounded above by $K + (K - 1) + \ldots + 1 = K(K + 1)/2$. •

The class of distribution functions with (7.21) is nonempty. For instance, it includes $F(x) = x^2$ with $\gamma \leq \sqrt{\sqrt{2} - 1}$, i.e., $\Delta \leq 0.388$. In the case of the uniform distribution ($F(x) = x$), condition (7.21) becomes an identity.

Apparently, a result similar to Assertion 7.6 takes place for the "concave" distribution functions (with inverse inequality of type (7.21)). However, for these functions the nonzero CBE is achievable for arbitrarily small control actions (see models in Chap. 5). In the sense of applications, this case seems less interesting.

Concluding Sect. 7.3, consider the "asymptotics" of the problems as $K = +\infty$. Suppose that (a) the threshold distribution function $F(\cdot)$ is continuous with a unique inflection point and $F(0) = 0$, (b) the equation $F(x) = x$ has a unique solution $q > 0$ on the segment $[0, 1]$ so that $\forall x \in (0; q): F(x) < x, \forall x \in (q; 1): F(x) > x$. Several examples of $F(\cdot)$ satisfying these assumptions are the functions obtained by the identification of a series of real online social networks, see [13]. The Principal seeks to excite all agents with the minimum costs.

By the above assumptions on $F(\cdot)$, if for some step l we have $x_l > q$, then the sequence $x_k(\delta)$ is nondecreasing for all $k > l$, $\delta_k \equiv 0$ and, furthermore, $\lim\limits_{k \to +\infty} x_k = 1$. This property of the mob admits the following interpretation. *The domain of attraction* of the zero equilibrium without control (without introduced provokers) is the half-interval $[0; q)$. In other words, it takes the Principal only to excite more than the proportion q of the agents; subsequently, the mob itself surely converges to the unit equilibrium even without control.

Denote by δ^l the solution of the problem

$$\sum_{k=0}^{l-1} \frac{\delta_k}{1-\delta_k} \rightarrow \min_{\delta:\, \delta_k \in [0;\Delta],\, x_l(\delta) > q}. \tag{7.22}$$

Calculate $Q_l = \sum_{k=0}^{l-1} \frac{\delta_k^l}{1-\delta_k^l}$ and find $l^* = \arg\min_{l=1,2,\dots} Q_l$.

The solution of problem (7.22) exists under the condition

$$\forall x \in [0, q] \rightarrow \Delta + (1 - \Delta)F(x) > x,$$

which can be rewritten as

$$\Delta > \max_{x \in [0;q]} [x - (1 - \Delta)F(x)]. \tag{7.23}$$

Let Δ^* be the infimum of all Δ satisfying (7.27). In fact, the quantity Δ^* represents the minimum value of the proportion of provokers required for the mob to "cross" the unstable equilibrium q, moving subsequently to the stable equilibrium with all excited agents. This quantity depends on the shape of $F(x)$ on the segment $[0, q]$, representing a quantitative characteristic of the mob's small excitability under few provokers. In this sense, Δ^* reflects the mob "inertia."

Assertion 7.7. $\Delta^* = \max_{x \in [0,q]} \frac{x - F(x)}{1 - F(x)}$.

Proof of Assertion 7.7. Take an arbitrary value Δ satisfying (7.23), i.e., $\Delta + (1 - \Delta)F(x) - x > 0$. Rewrite this inequality in the form $(1 - F(x))\left(\Delta - \frac{x - F(x)}{1 - F(x)}\right) > 0$. Here the first factor is positive and therefore

$$\forall x \in [0, q] \rightarrow \Delta > \frac{x - F(x)}{1 - F(x)},$$

which implies

$$\Delta \geq \max_{x \in [0,q]} \frac{x - F(x)}{1 - F(x)}$$

In other words, the value $\max_{x \in [0,q]} \frac{x - F(x)}{1 - F(x)}$ is the lower bound of the values Δ satisfying (7.23). On the other hand, for all $\varepsilon > 0$ the value $\max_{x \in [0,q]} \frac{x - F(x)}{1 - F(x)} + \varepsilon$ is not the lower bound, since for the smaller value $\Delta = \max_{x \in [0,q]} \frac{x - F(x)}{1 - F(x)} + \frac{\varepsilon}{2}$ condition (7.23) holds:

$$(1 - F(x))\left(\Delta - \frac{x - F(x)}{1 - F(x)}\right) = (1 - F(x))\left(\max_{x\in[0,q]} \frac{x - F(x)}{1 - F(x)} + \frac{\varepsilon}{2} - \frac{x - F(x)}{1 - F(x)}\right)$$

$$\geq (1 - F(x))\frac{\varepsilon}{2} > 0 \quad \forall x \in [0, q].$$

And so, by the definition of infimum, $\Delta^* = \max_{x\in[0,q]} \frac{x - F(x)}{1 - F(x)}$.•

Owing to the above assumptions on the properties of the distribution function, the optimal solution to the problem is characterized as follows.

Assertion 7.8. If condition (7.23) holds, then

$$\delta^{l^*} = (\delta_0^{l^*} > 0; \delta_1^{l^*} > 0; \ldots; \delta_{l^*-1}^{l^*} > 0; 0; 0; \ldots).$$

Assertion 7.8 implies that, in terms of excitation costs minimization for the mob satisfying the above assumptions, the control action calculated by (7.23) must be nonzero only at the first l^* steps (including the zero step).

Example 7.4. The paper [13] and Sect. 7.3 constructed the two-parameter threshold distribution function with the parameters a and b describing in the best way the evolvement of active users in the Russian-language segments of online social networks *LiveJournal*, *FaceBook*, and *Twitter*. This function has the form (also, see expression (3.17))

$$F_{a,b}(x) = \frac{\text{arctg}(a(x-b)) + \text{acrtg}(ab)}{\text{arctg}(a(1-b)) + \text{acrtg}(ab)}, a \approx [7, 15], b \in [0, 1]. \tag{7.24}$$

Choose $a = 13$ that corresponds to *Facebook* and $b = 0.4$. In this case, $q \approx 0.375$ and $\Delta^* \approx 0.169$.•

Note that we have considered the controlled mob withdrawal from the attraction domain $[0, q)$ of the zero point. A mob can "self-excite" owing to the appropriate realizations of the random variables (agents' thresholds) with a theoretical distribution function $F(\cdot)$. The models describing such effects via the framework of large deviations were considered in [27, 80, 81].

7.4 Probabilistic Models of Provokers Detection

Let us modify the model in the following way. At step k, the Principal chooses some proportion of provokers δ_k, and then another subject (for convenience, called *Metaprincipal*) observes the current situation, detecting the presence of provokers with a probability $p(\delta_k)$. Here $p(\cdot) : [0; 1] \to [0; 1]$ is a nondecreasing function such that $p(0) = 0$. If the provokers are detected, the game "stops," and the proportion of active agents is zero at all the subsequent steps. If the Metaprincipal detects no provokers, the proportion of active agents is defined by (7.11), the Principal chooses δ_{k+1}, and so on.

The probability that the provokers are not detected till step K makes up

$$P_K(\delta) = \prod_{k=0}^{K-1} (1 - p(\delta_k)). \tag{7.25}$$

The control problem solved by the Principal admits, e.g., the following statement:

$$\begin{cases} x_K(\delta) \to \max_\delta, \\ P_K(\delta) \geq \hat{P}, \\ (11), \ \delta_k \in [0; \Delta], \end{cases} \tag{7.26}$$

where $\hat{P}$ is a given probability that at step K the mob contains active agents. Formally speaking, problem (7.25)–(7.26) represents an analog of problem (7.15) and can be solved numerically in each special case. However, in some situations, it is possible to say something about the structure of the optimal solution. For instance, the next assertion states that, if the agents' thresholds obey the uniform distribution, then the unified solution is optimal (in this solution, the proportion of provokers introduced at each step is fixed, see Assertion 7.4).

Assertion 7.9. If $F(x) = x$ and $p(\rho) = \rho$, then there exists the optimal unified solution ρ^* of problem (7.25)–(7.26):

$$\rho^* = \min \{\Delta; 1 - (\hat{P})^{1/K}\}. \tag{7.27}$$

Proof of Assertion 7.9. As $x_K(\delta)$ and $P_K(\delta)$ do not decrease and increase, respectively, in all components of the vector δ, in the optimal solution the constraint $P_K(\delta) \geq \hat{P}$ becomes an equality.

According to Assertion 7.4, for any sequence δ there exists an equivalent unified solution (in terms of the terminal proportion of active agents).

Assume that in the sequence δ there exists l such that $0 < l < K-1$ and $\delta_l \neq \delta_{l+1} \in [0; \Delta]$. Now, demonstrate that for $p(\rho) = \rho$ we have

$$P_K(\delta) = P_K(\delta_0, \ldots, \delta_{l-1}, \rho(\delta_l, \delta_{l+1}), \rho(\delta_l, \delta_{l+1}), \delta_{l+2}, \ldots, \delta_{K-1})$$

Really, using expression (7.25), it can be easily verified that $(1-\delta_l)(1-\delta_{l+1}) \equiv (1 - \rho(\delta_l, \delta_{l+1}))^2$.

Formula (7.27) follows from the unified character of the optimal solution and expression (7.25).•

And finally, note another probabilistic model of provokers detection that proceeds from the following assumption. Being detected by the Metaprincipal at step k, at the next step the provokers are eliminated (their number is nullified) and the "game" resumes for the Principal and the residual proportion of provokers ($\Delta - \delta_k$).

Consequently, this section has studied different mob excitation problems in discrete time based on Granovetter's model, on the finite and infinite horizons. For a series of problems with given mob characteristics, the optimal solutions (in a certain sense) have been obtained or reduced to well-known numerical optimization problems.

Chapter 8
Dynamic Models of Mob Control in Continuous Time

This section written jointly with I.N. Barabanov is dedicated to the continuous-time models of mob control [10]. We formulate and solve the mob excitation control problem in the continuous-time setting by introducing an appropriate number of "provokers" at each moment of control.

The mob model proper is imported from [1, 79], actually representing a generalization of Granovetter's model to the continuous-time case as follows. Suppose that we know the *proportion* $x_0 \in [0, 1]$ of active agents at the initial (zero) moment. Then the evolution of this proportion $x(t)$ in the continuous time $t \geq 0$ is governed by the equation

$$\dot{x} = F(x) - x, \tag{8.1}$$

where $F(\cdot)$ is a known continuous function possessing the properties of a distribution function, $F(0) = 0$ and $F(1) = 1$. Actually, this is the distribution function of the agents' thresholds. Just like in [25], as well as Chaps. 5 and 7 above, by applying a control action $u(t) \in [0, 1]$ (introducing provokers), we obtain the controlled dynamic system

$$\dot{x} = u(t) + (1 - u(t))F(x) - x. \tag{8.2}$$

This section is organized in the following way. Section 8.1 studies the attainability set and the monotonicity of the system trajectories in the control action. Next, Sect. 8.2 is focused on the case of constant controls according to the above classification. Section 8.3 considers the models where control excites the whole mob. And finally, Sect. 8.4 deals with the case of positional control.

© Springer International Publishing AG 2017
V.V. Breer et al., *Mob Control: Models of Threshold Collective Behavior*,
Studies in Systems, Decision and Control 85,
DOI 10.1007/978-3-319-51865-7_8

8.1   Attainability Set and Monotonicity

First, we formulate a lemma required for further analysis.

Consider functions $G_1(x,t)$ and $G_2(x,t)$: $R \times [t_0, +\infty) \to R$ that are continuously differentiable with respect to x and continuous in t. By assumption, the functions G_1 and G_2 are such that the solutions to the Cauchy problems for the differential equations $\dot{x} = G_i(x,t)$, $i = 1,2$, with initial conditions (t_0, x_0), $x_0 \in R$, admit infinite extension in t. Denote by $x_i(t, (t_0, x_0))$, $i = 1,2$, the solutions of the corresponding Cauchy problems.

Lemma 8.1 *Let* $\forall x \in R$, $\forall t \geq t_0 \to G_1(x,t) > G_2(x,t)$. *Then* $\forall t > t_0 \to x_1(t, (t_0, x_0)) > x_2(t, (t_0, x_0))$.

Proof of Lemma 8.1. By the hypothesis, $G_1(x_0, t_0) > G_2(x_0, t_0)$, i.e., $\frac{d}{dt} x_1(t, (t_0, x_0))\big|_{t=t_0} > \frac{d}{dt} x_2(t, (t_0, x_0))\big|_{t=t_0}$. And so, there exists a number $\varepsilon > 0$ such that $\forall t \in (t_0, t_0 + \varepsilon] \to x_1(t, (t_0, x_0)) > x_2(t, (t_0, x_0))$, implying that on the whole half-interval $(t_0, t_0 + \varepsilon]$ the curve of the solution to the first Cauchy problem lies higher than its counterpart in the second Cauchy problem. We will demonstrate that this geometric relationship holds for all $t \geq t_0$.

Prove by contradiction, supposing that $\exists \hat{t} : x_1(\hat{t}, (t_0, x_0)) = x_2(\hat{t}, (t_0, x_0)) = \hat{x}$. Without loss of generality, let $\hat{t}$ be the first moment when the curve x_2 reaches the curve x_1, i.e., $\hat{t} = \inf\{t > t_0 + \varepsilon : x_1(t, (t_0, x_0)) = x_2(t, (t_0, x_0))\} < +\infty$. Obviously, $\hat{t} \geq t_0 + \varepsilon > t_0$ and $\forall t \in (t_0, \hat{t}) \to x_1(t, (t_0, x_0)) > x_2(t, (t_0, x_0))$. Hence, for $\tau \in [0, \hat{t} - t_0)$ we have

$$x_1(\hat{t} - \tau, (t_0, x_0)) - x_1(\hat{t}, (t_0, x_0)) > x_2(\hat{t} - \tau, (t_0, x_0)) - x_2(\hat{t}, (t_0, x_0)),$$

since the second terms in both sides of this inequality coincide with $\hat{x}$. Divide both sides of the inequality by $-\tau$ (reversing its sign) and proceed to the limits as $\tau \to 0$. As a result, the derivatives of the solutions x_1 and x_2 at the point $\hat{t}$ satisfy $\frac{d}{dt} x_1(t, (t_0, x_0))\big|_{t=\hat{t}} \leq \frac{d}{dt} x_2(t, (t_0, x_0))\big|_{t=\hat{t}}$, whence it appears that $G_1(\hat{x}, \hat{t}) \leq G_2(\hat{x}, \hat{t})$. This inequality contradicts the condition of Lemma 8.1, and the conclusion follows.•

Note that, for validity of this lemma, one should not consider the inequality $G_1(x,t) > G_2(x,t)$ for all $x \in R$. It suffices to take the union of the attainability sets of the equations $\dot{x} = G_i(x,t)$, $i = 1,2$, with the chosen initial conditions (t_0, x_0).

Designate by $x_t(u)$ the proportion of active agents at the moment t under the control action $u(\cdot)$. The right-hand side of expression (8.1) increases monotonically in u for each t and $\forall x \in [0, 1]$: $F(x) \leq 1$. Hence, we have the following result.

Assertion 8.1 *Let the function $F(x)$ be such that $F(x) < 1$ for $x \in [0, 1)$. If $\forall t \geq t_0 \to u_1(t) > u_2(t)$ and $x_0(u_1) = x_0(u_2)$ ($x_0 < 1$), then $\forall t > t_0$: $x_t(u_1) > x_t(u_2)$.*

Really, by the premises, for all t and $x < 1$ we have the inequality

$$u_1(t) + (1 - u_1(t))F(x) - x > u_2(t) + (1 - u_2(t))F(x) - x,$$

as the convex combination of different numbers (1 and $F(x)$) is strictly monotonic. The point $x = 1$ forms the equilibrium of system (8.1) under any control actions u (t). And so, it is unattainable for any finite t. Using Lemma 8.1, we obtain $x_t(u_1) > x_t(u_2)$ under the same initial conditions.

Suppose that the control actions are subjected to the *constraint*

$$u(t) \leq \Delta, t \geq t_0, \tag{8.3}$$

where $\Delta \in [0, 1]$ means some constant.

We believe that $t_0 = 0$, $x(t_0) = x(0) = 0$, i.e., initially the mob is in the nonexcited state.

If the *efficiency criterion* is defined as the proportion of active agents at a given moment $T > 0$, then the corresponding *terminal control problem* takes the form

$$\begin{cases} x_T(u) \to \max_{u(\cdot)}, \\ (8.2), \ (8.3). \end{cases} \tag{8.4}$$

Here is a series of results (Assertions 8.2–8.4) representing the analogs of the corresponding assertions from Chap. 7.

Assertion 8.2   The solution of problem (8.4) is given by $u(t) = \Delta$, $t \in [0, T]$.

Denote by $\tau(\hat{x}, u) = \min \{t \geq 0 | x_t(u) \geq \hat{x}\}$ the first moment when the proportion of active agents achieves a required value $\hat{x}$ (if the set $\{t \geq 0 | x_t(u) \geq \hat{x}\}$ is empty, just specify $\tau(\hat{x}, u) = +\infty$). Within the current model, one can pose the following *performance problem*:

$$\begin{cases} \tau(\hat{x}, u) \to \min_{u(\cdot)}, \\ (8.2), \ (8.3). \end{cases} \tag{8.5}$$

Assertion 8.3   The solution of problem (8.5) is given by $u(t) = \Delta$, $t \in [0, \tau]$.

By analogy with the discrete-time models (see Chap. 7), problem (8.4) or (8.5) has the following practical interpretation. The Principal benefits most from introducing the maximum admissible number of provokers in the mob at the initial moment, doing nothing after that (e.g., instead of first decreasing and then again increasing the number of introduced provokers). This structure of the optimal solution can be easily explained, as in models (8.4) and (8.5) the Principal incurs no costs to introduce and/or keep the provokers.

What are the properties of the *attainability set* $D = \bigcup_{u(t) \in [0; \Delta]} x_T(u)$? Clearly, $D \subseteq [0; 1]$, since the right-hand side of the dynamic system (8.2) vanishes for $x = 1$.

In the sense of potential applications, a major interest is attracted by the case of *constant controls* ($u(t) = v$, $t \geq 0$). Here the Principal chooses a same proportion

$v \in [0, \Delta]$ of provokers at all moments. Let $x_T(\Delta) = x_T(u(t) \equiv \Delta)$, $t \in [0, T]$, and designate by $D_0 = \bigcup_{v \in [0;\Delta]} x_T(v) \subseteq [0; 1]$ the attainability set under the constant control actions. According to Assertion 8.1, $x_T(v)$ represents a monotonic continuous mapping of $[0, \Delta]$ into $[0, 1]$ such that $x_T(0) = 0$. This leads to the following.

Assertion 8.4 $D_0 = [0, x_T(\Delta)]$.

Consider the models taking into account the *Principal's control costs*. Given a fixed "price" $\lambda \geq 0$ of one provoker per unit time, the Principal's costs over a period $\tau \geq 0$ are defined by

$$c_\tau(u) = \lambda \int_0^\tau u(t)dt. \qquad (8.6)$$

Suppose that we know a pair of monotonic functions characterizing the Principal's terminal payoff $H(\cdot)$ from the proportion of active agents and his current payoff $h(\cdot)$. Then problem (8.4) can be "generalized" to

$$\begin{cases} H(x_T(u)) + \int_0^T h(x(t))dt - c_T(u) \to \max\limits_u, \\ (8.2), (8.3). \end{cases} \qquad (8.7)$$

Under the existing constraints on the Principal's "total" costs C, problem (8.7) acquires the form

$$\begin{cases} H(x_T(u)) + \int_0^T h(x(t))dt \to \max\limits_u, \\ (8.2), c_T(u) \leq C. \end{cases} \qquad (8.8)$$

A possible modification of problems (8.4), (8.5), (8.7), and (8.8) is the one where the Principal achieves a required proportion $\hat{x}$ of active agents by the moment T (the *cost minimization problem*):

$$\begin{cases} c_T(u) \to \min\limits_u, \\ x_T(u) \geq \hat{x}, \\ (8.2). \end{cases} \qquad (8.9)$$

The problems of form (8.7)–(8.9) can be easily reduced to standard optimal control problems.

Example 8.1 Consider problem (8.9) where $F(x) = x$, $x_0 = 0$ and the Principal's costs are defined by (8.6) with $\lambda_0 = 1$. This yields the following optimal programmed control problem with fixed bounds:

$$\dot{x} = u(1 - x),$$
$$x(0) = 0, \ x(T) = \hat{x},$$
$$0 \le u \le \Delta, \tag{8.10}$$
$$\int_0^T u(t)dt \to \min_{u \in [0,\Delta]}.$$

For problem (8.10), construct the Pontryagin function $H = \psi(u(1 - x)) - u$. By the maximum principle, this function takes the maximum values in u. As the function enjoys linearity in u, its maximum is achieved at an end of the interval $[0, \Delta]$ depending on the sign of the factor at u, i.e.,

$$u = \frac{\Delta}{2}(sign(\psi(1 - x) - 1) + 1). \tag{8.11}$$

That the Pontryagin function is linear in the control actions actually follows from the same property of the right-hand side of the dynamic system (8.2) and functional (8.6). In other words, we have the result below.

Assertion 8.5 If the constraints in the optimal control problems (8.7)–(8.9) are linear in the control actions, then the optimal programmed control possesses the structure described by (8.11). That is, at each moment the control action takes either the maximum or minimum admissible value.

The Hamilton equations acquire the form

$$\dot{x} = \frac{\partial H}{\partial \psi} = u(1 - x), \ \dot{\psi} = -\frac{\partial H}{\partial x} = u\psi.$$

The boundary conditions are imposed on the first equation only. For $u = 0$, its solution is a constant; for $u = \Delta$, the function $x(t) = 1 - (1 - x(t_0))e^{-\Delta(t-t_0)}$.

The last expression restricts the maximum number of provokers required for mob transfer from the zero state to $\hat{x}$: $\Delta \ge \frac{1}{T}\log\frac{1}{1-\hat{x}}$.

And there exists the minimum time $t_{min} = \frac{1}{\Delta}\log\frac{1}{1-\hat{x}}$, during which the control actions take the maximum value Δ, being 0 at the rest moment. Particularly, a solution of problem (8.10) has the form

$$u = \begin{cases} \Delta, & t \le t_{min} \\ 0, & t_{min} < t \le T \end{cases}, \tag{8.12}$$

when the Principal introduces the maximum number of provokers from the very beginning, maintaining it during the time t_{min}.

The structure of the optimal solution to this problem (a piecewise constant function taking values 0 or Δ) possibly requires minimizing the number of control switchovers (discontinuity points). Such an additional constraint reflects the situations when the Principal incurs extra costs to introduce or withdraw provokers.

If this constraint appears in the problem, the best control actions in the optimal control set are either (8.12) or $u = \begin{cases} \Delta, & t \in [T - t_{min}, T], \\ 0, & t < T - t_{min}. \end{cases}$ •

8.2 Constant Control

In the class of the constant control actions, we obtain $c_\tau(v) = \lambda\, v\, \tau$ from formula (8.6). Under given functions $F(\cdot)$, i.e., a known relationship $x_t(v)$, problems (8.7)–(8.9) are reduced to standard scalar optimization problems.

Example 8.2   Choose $F(x) = x$, $T = 1$, $x_0 = 0$, $H(x) = x$, and $h(x) = \gamma x$, where $\gamma \geq 0$ is a known constant. It follows from (8.1) that

$$x_t(u) = 1 - \exp\left(-\int_0^t u(y)dy\right). \tag{8.13}$$

For the constant control actions, $x_t(v) = 1 - e^{-vt}$.
Problem (8.7) becomes the scalar optimization problem

$$e^{-v}\left(\frac{\gamma}{v} - 1\right) - \frac{\gamma}{v} - \lambda v \to \max_{v \in [0;\Delta]}. \tag{8.14}$$

Next, problem (8.8) becomes the scalar optimization problem

$$e^{-v}\left(\frac{\gamma}{v} - 1\right) - \frac{\gamma}{v} \to \max_{v \in [0;\Delta]}. \tag{8.15}$$

And finally, problem (8.9) acquires the form $\begin{cases} v \to \min_{v \in [0;1]}, \\ 1 - e^{-v} = \hat{x}. \end{cases}$ Its solution is described by $v = \ln\left(\frac{1}{1-\hat{x}}\right)$. •

8.3 Excitation of Whole Mob

Consider the "asymptotics" of the problems as $T \to +\infty$. Similarly to the corresponding model in Chap. 7, suppose that (a) the function $F(\cdot)$ has a unique inflection point and $F(0) = 0$, (b) the equation $F(x) = x$ has a unique solution $q > 0$ on the interval $(0, 1)$ so that $\forall x \in (0; q): F(x) < x; \forall x \in (q; 1): F(x) > x$. Several examples of the functions $F(\cdot)$ satisfying these assumptions are provided in Chap. 3. The Principal seeks to excite all agents with the minimum costs.

By the above assumptions on $F(\cdot)$, if for some moment τ we have $x(\tau) > q$, then the trajectory $x_t(u)$ is nonincreasing and $\lim_{t \to +\infty} x_t(u) = 1$ even under $u(t) \equiv 0$ $\forall t > \tau$. This property of the mob admits the following interpretation. The *domain of attraction* of the zero equilibrium without control (without the introduced provokers) is the half-interval $[0, q)$. In other words, it takes the Principal only to excite more than the proportion q of the agents; subsequently, the mob itself surely converges to the unit equilibrium even without control.

Denote by u^τ the solution of the problem

$$\int_0^\tau u(t)dt \;\longrightarrow\; \min_{u:\, u(t) \in [0;\Delta],\, x_\tau(u) > q} . \tag{8.16}$$

Calculate $Q_\tau = \int_0^\tau u^\tau(t)dt$ and find $\tau^* = \arg\min_{\tau \geq 0} Q_\tau$.

The solution of problem (8.16) exists under the condition

$$\Delta > \Delta^* = \max_{x \in [0,q]} \frac{x - F(x)}{1 - F(x)}. \tag{8.17}$$

For practical interpretations, see the previous sections.

Owing to the above assumptions on the properties of the distribution function, the optimal solution to the problem is characterized as follows.

Assertion 8.6 If condition (8.17) holds, then $u^\tau(t) \equiv 0$ for $t > \tau$.

Example 8.3 The paper [13] (see also Chap. 3) constructed the two-parameter function $F(\cdot)$ describing in the best way the evolvement of active users in the Russian-language segments of online social networks *LiveJournal*, *FaceBook* and *Twitter*. The role of the parameters is player by a и b. This function has the form

$$F_{a,b}(x) = \frac{arctg(a(x - b)) + acrtg(ab)}{arctg(a(1 - b)) + acrtg(ab)}, \quad a \approx [7;\, 15],\; b \in [0;\, 1]. \tag{8.18}$$

Choose $a = 13$ that corresponds to *Facebook* and $b = 0.4$. In this case, $q \approx 0.375$ and $\Delta^* \approx 0.169$; the details can be found in [9].•

8.4 Positional Control

In the previous subsections, we have considered the optimal programmed control problem arising in mob excitation. An alternative approach is to use positional control. Consider two possible statements having transparent practical interpretations.

Within the first statement, the problem is to find a positional control law $\tilde{u}(x)$: $[0, 1] \rightarrow [0, 1]$ ensuring maximum mob excitation (in the sense of (8.4) or (8.5)) under certain constraints imposed on the system trajectory and/or the control actions.

By analogy with expression (8.3), suppose that the control actions are bounded:

$$\tilde{u}(x) \leq \Delta, \quad x \in [0, 1], \tag{8.19}$$

and there exists an additional constraint on the system trajectory in the form

$$\dot{x}(t) \leq \delta, \quad t \geq 0, \tag{8.20}$$

where $\delta > 0$ is a known constant. Condition (8.20) means that, e.g., a very fast growth of the proportion of excited agents (increment per unit time) is detected by appropriate authorities banning further control. Hence, trying to control mob excitation, the Principal has to maximize the proportion of excited agents subject to conditions (8.19) and (8.20). The corresponding problem possesses the simple solution

$$\tilde{u}^*(x) = \min \left\{ \Delta; \; \max \left\{ 0; \; \frac{x + \delta - F(x)}{1 - F(x)} \right\} \right\}. \tag{8.21}$$

owing to the properties of the dynamic system (8.2), see Lemma 8.1. The fraction in (8.21) results from making the right-hand side of (8.1) equal to the constant δ.

Note that, under small values of δ, the nonnegative control action satisfying (8.20) may cease to exist.

Example 8.4 Within the conditions of Example 8.3, choose $\delta = 0.35$. Then the optimal positional control is illustrated by Fig. 8.1 (dashed lines indicate the bisecting line (blue color) and the curve $F(\cdot)$ (red color), while thin solid line shows the upper bound appearing in (8.21)).•

The second statement of positional control relates to the so-called *network immunization problem*, see Chap. 5. Here the Principal seeks to reduce the proportion of active agents by introducing an appropriate number (or proportion) of *immunizers*–agents that always prefer passivity.

Denote by $w \in [0, 1]$ the proportion of immunizers. As shown above, the proportion of active agents evolves according to the equation

$$\dot{x} = (1 - w)F(x) - x, \quad x \in [0, 1). \tag{8.22}$$

Let $\tilde{w}(x) : [0, 1] \rightarrow [0, 1]$ be a positional control action. If the Principal is interested in reducing the proportion of active agents, i.e.,

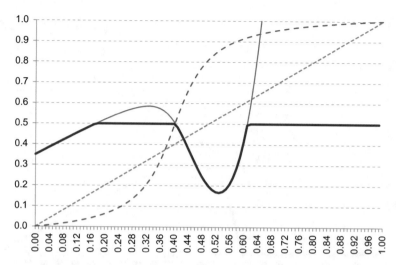

Fig. 8.1 Optimal positional control in Example 8.4

$$\dot{x}(t) \leq 0, \quad t \geq 0, \tag{8.23}$$

then the control actions must satisfy the inequality

$$\tilde{w}(x) \geq 1 - \frac{x}{F(x)}. \tag{8.24}$$

The quantity $\Delta_{min} = \max\limits_{x \in [0;1]} (1 - \frac{x}{F(x)})$ characterizes the minimal restrictions on the control actions at each moment when system (8.22) is "controllable" in the sense of (8.23).

Example 8.5 Within the conditions of Example 8.4, the lower bound (8.24) of the positional control is shown by Fig. 8.2, see thick line. Again, dashed lines indicate the bisecting line (blue color) and the distribution function $F(\cdot)$.

Here the quantity Δ_{min} is approximately 0.385.•

This section has described the continuous-time problems of mob excitation control using the introduction of provokers or immunizers.

A promising line of future research is the analysis of a differential game describing the informational confrontation of two control subjects (Principals) that choose in continuous time the proportions (or numbers) of the introduced provokers u and immunizers w, respectively. The corresponding static problem [79] can be a "reference model" here. The controlled object is defined by the dynamic system

$$\dot{x} = u(1 - w) + (1 - u - w + 2uw)F(x) - x.$$

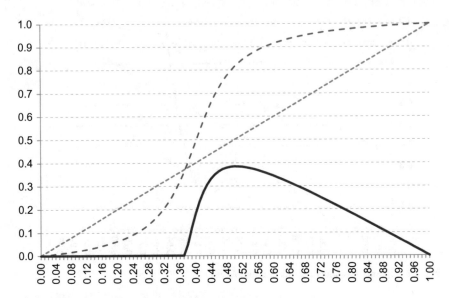

Fig. 8.2 Minimal positional control in Example 8.5

Another line of interesting investigations concerns the mob excitation problems with dynamic (programmed and/or positional) control, where the mob obeys the transfer equation of the form

$$\frac{\partial}{\partial t}p(x,t) + \frac{\partial}{\partial x}\left(\left[u + (1-u)F(x) - x\right]p(x,t)\right) = 0. \tag{8.25}$$

In this model, at each moment the mob state is described by a probability distribution function $p(x, t)$, instead of the scalar proportion of active agents.

Chapter 9
Micromodels of Informational Confrontation

To characterize the informational confrontation problems solved at level 5 of the ANS description (see the Introduction), one needs simple results in the fields of informational interaction analysis and informational control of the agents.

The first class of the models with a complete chain between the lower and upper levels was the models of ANSs described in terms of the micromodel, i.e., the "consensus problems" (or the so-called Markov models). As a matter of fact, this advancement facilitated the development of the corresponding game-theoretic models of informational confrontation [46] considered in this section. Within these models, the players exert certain impact on the elements of an ANS, being interested in its specific states. And the ANS proper is described by the micromodel taking into account the individual actions of the agents.

The second fruitful example is the design approach to the game-theoretic models of informational confrontation that are "superstructed" over the threshold macromodels of a mob, the approach cultivated in the next section.

A common feature of both approaches consists in a natural aspiration for "superstructing" a game-theoretic model (characterizing the interaction among the control subjects) over the model of an ANS. The current section is intended to demonstrate the feasibility of such symbiosis (more specifically, the feasibility of reducing the problems of informational control analysis in the ANSs to some classical problems of game theory). Further exposition has the following structure. Section 9.1 describes the model of an ANS, the model of informational control and game-theoretic model of informational confrontation, respectively. Then, Sects. 9.2–9.5 provide illustrative examples with different concepts of game equilibrium, viz., dominant strategies' equilibrium, Nash equilibrium, "contract equilibrium" (being Pareto efficient), Stackelberg equilibrium, informational equilibrium, and secure strategies' equilibrium.

© Springer International Publishing AG 2017
V.V. Breer et al., *Mob Control: Models of Threshold Collective Behavior*,
Studies in Systems, Decision and Control 85,
DOI 10.1007/978-3-319-51865-7_9

9.1   Micromodel and Informational Control

Consider an ANS composed of n *agents*. The *opinion* of agent i at step t represents a real value x_i^t, $i \in N = \{1, 2, \ldots, n\}$, $t = 0, 1, 2, \ldots$. Following the tradition of Markov models [31, 33, 48, 78], let us reflect the mutual informational impact of the agents by a nonnegative *confidence matrix* $A = \|a_{ij}\|$ being stochastic in rows; here a_{ij} indicates the confidence level of agent i in agent j $(i, j \in N)$. Suppose that the vector $x = (x_i^0) i \in N$ of the initial opinions of the agents is given. Moreover, at each step agent i changes his opinion considering the opinions of those agents he trusts (including his own opinion):

$$x_i^t = \sum_{j \in N} a_{ij} x_j^{t-1}, t = 1, 2, \ldots, i \in N. \tag{9.1}$$

Assume that the agents interact for a sufficiently long time so that the vector of the final ("equilibrium") opinions is estimated by

$$x = A^{\infty} x^0, \tag{9.2}$$

where $A = \left[\lim_{t \to \infty} (A)^t \right]$. The corresponding convergence conditions can be found in [31, 78].

In the sequel, we will believe that each agent trusts (at least, a bit) the rest agents, that is, $a_{ij} > 0$, $i, j \in N$. As shown in [48], under this assumption.

- all rows of the matrix A^{∞} are identical (denote $r_j = a_{ij}^{\infty} > 0, i, j \in N$), while the matrix elements can be interpreted as the *impact levels* of the agents;
- the final opinions of all agents coincide (denote $X = x_i$, $i \in N$, $X \in \mathfrak{R}^1$),

  i.e., expression (9.2) takes the form

$$X = \sum_{j \in N} r_j x_j^0. \tag{9.3}$$

Note that similar result holds for the models of social networks where the confidence level is defined by the reputation of the agents, see above.

Among other things, *informational control* in the ANSs lies in a purposeful impact exerted on the initial opinions of agents, in order to ensure the required values of the final opinions (desired by the control subject).

Consider two players being able to influence the initial opinions of some agents. Let $F \subseteq N$ be the set of agents whose opinions are formed by the first player (the *impact agents* of the first player), and $S \subseteq N$ be the set of the impact agents of the second agent, where $F \cap S = \emptyset$. Assume that informational control is unified [74] in the following sense: the initial opinion $u \in U$ ($v \in V$) is formed for all agents belonging to the set F (S, respectively), with U and V representing intervals on $\mathfrak{R}^1$.

Denote $r_F := \sum_{i \in F} r_i, r_S := \sum_{j \in S} r_j, X := \sum_{k \in N \setminus (F \cup S)} r_k x_k^0$; then formula (9.3) is reduced to

$$X(u, v) = r_F u + r_S v + X^0. \tag{9.4}$$

Hence, the final opinion of the ANS members depends linearly on the control actions u and v with the weights $r_F > 0$ and $r_S > 0$, respectively, where $r_F + r_S \leq 1$. These weights are determined by the total impact level of the impact agents.

Making a digression, note that a different situation is when the impact agents fix their opinions: $a_{ij} = 0, j \neq i, i \in F \cup S$. Also, see the influence models of mass media in [48].

Using relationship (9.4) between the final opinion of the agents and the control actions, one can suggest a game-theoretic interaction model of the agents performing these actions. To this end, it is necessary to define their goal functions. Suppose that the goal functions of the first and second agents, $f_F(u, v) = H_F(X(u, v)) - c_F(u)$ and $f_S(u, v) = H_S(X(u, v)) - c_S(v)$, are calculated as the difference between their "income" that depends on the final opinion of the agents and the control costs.

The aggregate $\Gamma = \{f_F(u, v), f_S(u, v), u \in U, v \in V\}$ composed of the goal functions and feasible action sets of the two agents specifies the family of *games*. The distinctions among these games are induced by the awareness structure of the players and the sequence of functioning (see [74]).

Imagine that the description of the game Γ and formula (9.4) are the common knowledge of the agents choosing their actions one-time, simultaneously and independently. In this case, a *normal-form game* arises naturally, and further analysis can be focused on calculation of the corresponding *Nash equilibria*, their efficiency in the Pareto sense, etc. By fixing the sequence of actions choice for the players, one obtains a certain hierarchical game. On the other hand, rejecting the hypothesis of common knowledge yields a reflexive game [75], and so on—see several special cases below. With just a few exceptions, their practical interpretations are omitted due to obviousness. Many examples can be found in [48], as well.

9.2   "Antagonistic" Game

Choose the zero opinions of the agents ($X^0 = 0$) as the "status quo." Suppose that the first player is interested in final opinion maximization ($H_F(X) = X$), whereas the second player seeks to minimize it ($H_F(X) = -X$). Both players have identical "control resources" ($U = V = [d, D], d < -1 \leq 1 < D$) and identical cost functions ($c_F(u) = u^2/2, c_S(v) = v^2/2$).

The goal functions of the players,

$$f_F(u, v) = r_F u + r_S v - u^2/2 \qquad (9.5)$$

and

$$f_S(u, v) = -r_F u - r_S v - v^2/2, \qquad (9.6)$$

are separable in the corresponding actions. Hence [74], under the simultaneous independent choice of the players' actions, there exists a *dominant strategies' equilibrium* (DSE) (u^d, v^d), where $u^d = r_F$ and $v^d = -r_S$.

A *Pareto point* is the vector (u^P, v^P) that maximizes the sum of the goal functions of the players, where $u^P = 0$ and $v^P = 0$.

The DSE is Pareto inefficient:

$$f_F(u^d, v^d) + f_S(u^d, v^d) = -[(r_F)^2 + (r_S)^2]/2 < f_F(u^P, v^P) + f_S(u^P, v^P) = 0,$$

while the Pareto point appears unstable against the unilateral deviations of the players.

For the first (second) player, define the penalty strategy as his worst action for the opponent: $u^p = D$, $v^p = d$. Within the framework of this model, the dominant strategies of the players are *guaranteeing*. Calculate the guaranteed payoffs of the players:

$$f_F^{MGR} = f_F(u^d, v^p) = (r_F)^2/2 + r_S d, \quad f_S^{MGR} = f_S(u^p, v^d) = (r_S)^2/2 - r_F D.$$

Assume that a third party controls how the players fulfill their commitments [59, 74] and the following *contracts* are concluded (the "non-aggression pact"):

$$\hat{u}(v) = \begin{cases} 0, & v = 0 \\ u^p, & v \neq 0 \end{cases}, \quad \hat{v}(u) = \begin{cases} 0, & u = 0 \\ v^p, & u \neq 0 \end{cases}. \qquad (9.7)$$

Then the players benefit from executing these contracts if

$$\begin{cases} (r_F)^2 + 2 r_S d \leq 0, \\ (r_S)^2 \leq 2 r_F D, \end{cases} \qquad (9.8)$$

which leads to the stable implementation of the Pareto point. The same result can be achieved using the penalty strategy in *repeated games*. According to condition (9.8), the "*contract equilibrium*" is implementable if the impact levels of the first and second players differ slightly. Figure 9.1 demonstrates the hatched domain 0AB that satisfies condition (9.8) with $d = -1$ and $D = 1$.

Fig. 9.1 The "weights" of impact agents ensuring the existence of "contract" equilibrium

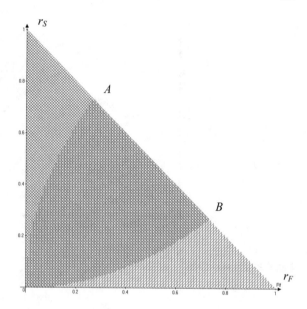

9.3 "Nonantagonistic" Game

Consider a normal form game differing from the one described in the previous subsection only in the "income" functions of the agents. That is, let $H_F(X) = X - 2X^2$ and $H_S(X) = X - X^2$, meaning that the first and second players seek for obtaining the final opinions $X_F = 0.25$ and $X_S = 0.5$, respectively.

The goal functions of the players,

$$f_F(u, v) = (r_F u + r_S v) - (r_F u + r_S v)^2 - u^2/2 \qquad (9.9)$$

and

$$f_S(u, v) = (r_F u + r_S v) - (r_F u + r_S v)^2 - v^2/2, \qquad (9.10)$$

are no more separable in the corresponding actions. Therefore, we evaluate *Nash equilibrium*

$$u^* = \frac{r_F - 2r_F(r_S)^2}{4(r_F)^2 + 2(r_S)^2 + 1}, \quad v^* = \frac{r_S + 2r_S(r_F)^2}{4(r_F)^2 + 2(r_S)^2 + 1}. \qquad (9.11)$$

Figure 9.2 shows the parametric set of the Nash equilibria for $r_F = 0.1$ and $r_S \in [0, 0.9]$.

The relationships between the equilibrium actions of the players and the proportions of the impact agents (equivalently, the total reputations r_S and r_F of the impact agents) are shown by Fig. 9.3. The left-hand graph characterizes the general

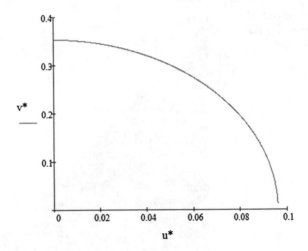

Fig. 9.2 The set of Nash equilibria

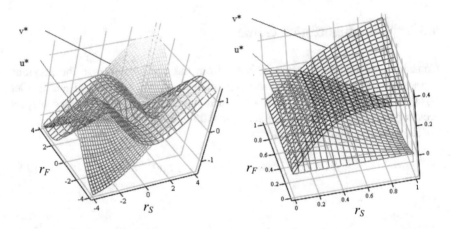

Fig. 9.3 3D graphs of u^* and v^*

form of this relationship, while the right-hand counterpart illustrates the relationship on the admissible ranges of r_S and r_F. Particularly, it can be observed that, the greater is the total reputation of the impact agents of the second player, the smaller is the equilibrium control action of the first player (on the one part) and the larger is the equilibrium control action of the second player (on the other part). That is, the first (second) player seeks for obtaining the smaller (greater, respectively) opinion.

Now, consider a hierarchical game of the type Γ_1 where the goal functions of the players are defined by (9.9) and (9.10) and the first player moves first.

Let us analyze the move made by the second player. At this step, the second player knows the opponent's action u and maximizes his own payoff by choosing the action $\frac{r_S - 2r_S r_F u}{2(r_S)^2 + 1}$. Actually, the choice set of the second player represents a singleton. The guaranteeing strategy of the first player in the game Γ_1 and his strategy in the Stackelberg equilibrium are defined by

$$u^* = \frac{r_F - 2r_F(r_S)^2}{4(r_F)^2 + 4(r_S)^4 + 4(r_S)^2 + 1}; \quad v^* = \frac{2(r_S)^3 - (2(r_F)^2 + 1)r_S}{4(r_F)^2 + 4(r_S)^4 + 4(r_S)^2 + 1}.$$

9.4 Reflexive Game

Consider the goal functions that differ from the ones explored in the previous subsection in the cost functions of the players: $c_F(u) = u^2/(2q_F)$, and $c_S(v) = v^2/(2q_S)$, where $q_F = 1$ and denote the "efficiency levels" of the players. By assumption, each player knows his efficiency level, the first player believes that the common knowledge is $q_S = 1$, while the second player knows this fact and the real efficiency of the first player. The described reflexive game [75] has the graph $2 \leftarrow 1 \leftrightarrow 12$.

According to expression (9.11), the first player chooses $u^* = \frac{r_F - 2r_F(r_S)^2}{4(r_F)^2 + 2(r_S)^2 + 1}$. Consequently, the second agent selects his best response $v^* = \frac{0.5r_S(1 + 2(r_F)^2)(1 + 2(r_S)^2)}{(1 + (r_S)^2)(4(r_F)^2 + 2(r_S)^2 + 1)}$. These actions lead to the final opinion $X = \frac{(r_F)^2 + (r_S)^4 + 0.5(r_S)^2}{(1 + (r_S)^2)(4(r_F)^2 + 2(r_S)^2 + 1)}$ of the ANS; in the general case, it does not coincide with the opinion $X^1 = \frac{(r_F)^2 + (r_S)^2}{4(r_F)^2 + 2(r_S)^2 + 1}$ expected by the first player. This means that the resulting informational equilibrium is unstable [75]. For this reflexive game graph, the informational equilibrium enjoys stability only in two situations: (1) the belief of the first agent about the opponent's efficiency level is true, or (2) the total reputation of the impact agents of the second player makes up zero (however, the reputation is assumed strictly positive).

9.5 Secure Strategies Equilibrium

Consider the game where $H_F(X(u, v)) = \begin{cases} h_F > 0, X \geq \hat{X} \\ 0, X < \hat{X} \end{cases}$, $H_S(X(u, v)) = \begin{cases} h_S > 0, X < \hat{X} \\ 0, X \geq \hat{X} \end{cases}$,

$c_F(u) = u$, and $c_S(v) = v$, with $U = V = [d, D]$, $d < -1 \leq 1 < D$, $h_F > D$, and $h_v > |d|$. In a corresponding practical interpretation, the first player is interested in the adoption of a certain decision, which requires that the opinion of the ANS members exceeds the threshold $\hat{X}$; in contrast, the second player seeks for blocking this decision.

Let $r_F D + r_S d + X^0 > \hat{X}$ for definiteness. There exists no Nash equilibria in this game, but it is possible to evaluate a *secure strategies' equilibrium* (SSE) [50, 51, 52] in the form $\left(\left(\hat{X} - r_S d - X^0 \right) / r_F + e; 0 \right)$ where ε is an arbitrary small strictly positive constant. A practical interpretation of the SSE lies in the following. The first player ensures the adoption of the required decision, and the second player (even choosing the maximum possible absolute values of the actions) is unable to modify the result.

In this section, we have analyzed a series of specific examples illustrating the applicability of game theory to the description of informational confrontation in ANSs, both in terms of the process and result. Despite its simplicity, the stated model shows the diversity of possible game-theoretic formulations (dominant strategies' equilibrium, Nash equilibrium, "contract equilibrium," Stackelberg hierarchical games and hierarchical games of the type Γ_1, reflexive games, secure strategies' equilibrium). Generally speaking, any specific model of informational confrontation should be developed taking into account, first, the features of an associated practical problem and, second, the identifiability of a modeled system (i.e., ANS parameters, the admissible actions of the players, as well as their preferences and awareness).

Note that the game-theoretic models of informational confrontation over ANSs have several applications, namely, information security in telecommunication networks, counteraction to the destructive informational impacts on the social groups of different scale, prevention of their massive illegal actions, and others.

As for the promising directions of further investigations, we mention the design and study of the game-theoretic models of informational confrontation in ANSs under the following conditions:

- the sequential choice of the players' actions under the observed dynamics of the ANS state (the "defense-attack" games with the description of opinion dynamics (the distribution of informational epidemic or viruses within an ANS));
- the repeated choice of the players' actions under incomplete information on the actions of the opponents and the ANS state.

Chapter 10
Macromodels of Informational Confrontation

Within the stochastic models of mob control (see Chap. 5), we explore the game-theoretic models of informational confrontation when the agents are simultaneously controlled by two subjects with noncoinciding interests regarding the number of active agents in an equilibrium state. And the ANS is described by the macromodel with the proportion of active agents as the main parameter.

This section is organized as follows. In Sect. 10.1, we treat informational confrontation within the stochastic models of mob control. And Sects. 10.2–10.5 provide original analysis results for the game-theoretic models of informational confrontation in terms of the normal-form games (including a characterization of Nash equilibria and secure strategies' equilibria), hierarchical games and reflexive games. Numerous examples show in the analytical form how these equilibria depend on the model parameters.

10.1 Model of Informational Confrontation

Consider a mob as an object controlled by two subjects, i.e., *Principals*. The behavior of the dynamic system (2.7) describing the evolution of the proportion of active agents is determined by the distribution function $F(\cdot)$ of the agents' thresholds. And so, we will analyze the control actions that change this distribution function.

Note that Chaps. 4 and 5 have defined the set (proportion) of the initially excited agents or/and the distribution function of their thresholds that implement a required equilibrium. Within the framework of the models studied below, the agents are excited "independently."

Designate by $\Lambda(x) = \{(\delta, \gamma) \in \mathfrak{R}^2_+ \,|\, x^*(\delta, \gamma) = x\}$ the set of the control pairs implementing a given value $x \in [0, 1]$ as the CBE, see Model II in Chap. 5.

© Springer International Publishing AG 2017
V.V. Breer et al., *Mob Control: Models of Threshold Collective Behavior*,
Studies in Systems, Decision and Control 85,
DOI 10.1007/978-3-319-51865-7_10

To explore the game-theoretic models of interaction between the Principals, we need a result that is proved similarly to Assertions 3 and 4 in the paper [25].

Assertion 10.1 In Model II, the CBE $x^*(\delta, \gamma)$ has the following properties:

(1) monotonic (nonstrict) increase in δ; for strict monotonicity, a sufficient condition is $F(1-0) < 1$ or $\gamma > 0$;
(2) monotonic (nonstrict) decrease in γ; for strict monotonicity, a sufficient condition is $F(0) > 0$ or $\delta > 0$.

Example 10.1 Consider the uniform distribution of the agents' thresholds, i.e., $F(x) = x$. Here $x^*(\delta, \gamma) = \delta/(\delta + \gamma)$ and $\Lambda(x) = \{(\delta, \gamma) \in \Re^2_+ \,|\, \gamma/\delta = (1/x - 1)\}$.•

As a digression, note an important feature of the socioeconomic and organizational systems with several subjects interested in certain states of a controlled system (e.g., a network of interacting agents) and applying control actions to it (*systems with distributed control* [43, 59, 74]). In such systems, just like in our case, there exists an interaction between the subjects, which is termed *informational confrontation* when they exert informational impacts on the controlled object.

In what follows, we consider a series of the game-theoretic models of interaction between the Principals whose informational impacts on a mob are defined by expression (5.10) (in Model I) or by expression (5.19) (in Model II).

10.2   Normal-Form Game

Model I. Two Principals exert an informational impact on a mob by playing a *normal-form game*. That is, Principal 1 and 2 choose their strategies $\alpha \in [0, 1]$ and $\beta \in [0, 1]$, respectively, one-time, simultaneously and independently from each other. The *goal functions* of Principals 1 and 2 have the form

$$f_\alpha(\alpha, \beta) = H_\alpha(x^*(\alpha, \beta)) - c_\alpha(\alpha), \tag{10.1}$$

$$f_\beta(\alpha, \beta) = H_\beta(x^*(\alpha, \beta)) - c_\beta(\beta), \tag{10.2}$$

Moreover, the *payoff* $H_\alpha(\cdot)$ of Principal 1 is an increasing function, as he seeks for maximizing the number of the excited agents, while the payoff $H_\beta(\cdot)$ of Principal 2 is a decreasing function (his interests are quite the opposite). Both *cost functions* $c_\alpha(\cdot)$ and $c_\beta(\cdot)$ are strictly increasing and $c_\alpha(0) = c_\beta = 0$.

The described game belongs to the class of the normal-form ones, and several typical questions of game theory arise immediately. What is the *Nash equilibrium* (α^*, β^*) in the agents' game? For which strategy profiles is the Nash equilibrium dominating the *status quo profile*, i.e., the CBE without control (i.e., when does the condition $f_\alpha(\alpha^*, \beta^*) \geq f_\alpha(0, 0)$, $f_\beta(\alpha^*, \beta^*) \geq f_\beta(0, 0)$ hold)? What is the structure of the set of Pareto efficient strategy profiles? When does a *dominant strategy equilibrium* (DSE) exist? And so on.

Denote by $f(\alpha, \beta) = f_\alpha(\alpha, \beta) + f_\beta(\alpha, \beta)$ the utilitarian collective utility function (CUF) [62]. The pair of strategies $(\hat{\alpha}; \hat{\beta}) = \arg \max\limits_{(\alpha,\beta) \in [0;1]^2} f(\alpha, \beta)$ will be called the *utilitarian solution*.

The results obtained in Chap. 5 and Assertion 10.1 are crucial for our game-theoretic analysis, as explained below. The goal functions (10.1) and (10.2) of the Principals depend on their strategies (α and β, or δ and γ) and on the CBE, which is by-turn dependent on these strategies. The monotonic dependence of the CBE on the strategies of the Principals (if necessary, its continuity can be checked in a specific case), as well as the implementability of the whole unit segment as the CBE via the appropriately chosen strategies allow "translating" the properties of the goal functions and cost functions on the dependence of these parameters directly from the strategies of the Principals. For instance, if $H_\delta(x^*(\delta, \gamma))$ is an increasing function of x^*, then by Assertion 10.1 the payoff of Principal 1 is an increasing function of his strategy, and so on.

An elementary case is the *antagonistic game*: Principal 1 seeks for maximizing the number of the excited agents, while Principal 2 pursues the opposite interests. Assuming $c_\alpha(\cdot) \equiv 0$ and $c_\beta(\cdot) \equiv 0$ (no control costs), the expressions (10.1) and (10.2) yield

$$\hat{f}_\alpha(\alpha, \beta) = x^*(\alpha, \beta),\ \hat{f}_\beta(\alpha, \beta) = 1 - x^*(\alpha, \beta). \tag{10.3}$$

Clearly, $f(\alpha, \beta) \equiv 1$. As $x^*(\alpha, \beta)$ does not decrease in α and increase in β, we naturally arrive at Assertion 10.2. Just like its "analogs" for Model II (see Assertions 10.3 and 10.4 below), this result seems trivial in some sense, following directly from the monotonicity of the goal functions of the agents in their actions. On the other hand, Assertion 10.2 proves the existence of the DSE and assists in its calculation in the degenerate cases.

Assertion 10.2 In Model I described by the antagonistic game with the zero control costs, there exists the DSE $\alpha^{DSE} = 1$, $\beta^{DSE} = 1$.

Note that, in this equilibrium, the distribution function of the agents' thresholds coincides with the initial distribution function, i.e., $F_{1,1}(x) \equiv F(x)$. Hence, the CBE remains invariable, "matching" the status quo profile.

Example 10.2 Choose $F(x) = x$, then

$$x^{I*}(\alpha, \beta) = \frac{\alpha(1 - \beta)}{\alpha + \beta - 2\alpha\beta}. \tag{10.4}$$

Evaluate

$$\frac{\partial x^{I*}(\alpha, \beta)}{\partial \alpha} = \frac{\beta(1 - \beta)}{(\alpha + \beta - 2\alpha\beta)^2}, \frac{\partial x^{I*}(\alpha, \beta)}{\partial \beta} = -\frac{\alpha(1 - \alpha)}{(\alpha + \beta - 2\alpha\beta)^2},$$

which shows that $x^{I*}(\alpha, \beta)$ is increasing in the first argument and decreasing in the second argument for any admissible values of the other argument. Therefore, under the zero control costs, the DSE in the Principals' game with the goal functions (10.3) is the unit strategies $\alpha^{DSE} = 1$, $\beta^{DSE} = 1$. Naturally, this point also represents the Nash equilibrium (NE) in the Principals' game. In the current example, we have $W = [0, 1]$. The DSE implements the same mob state as in the absence of control.•

Now, consider the case of nonzero control costs.

Assertion 10.3 In Model I with $W = [0, 1]$, $F(0) > 0$ and $F(1 - 0) < 1$, let $x^*(\alpha, \beta)$ be a continuous function, the payoff functions of the Principals be bounded, linear or concave in their strategies and the cost functions be convex. Then there exists a Nash equilibrium in the Principals' game.

The trivial Assertion 10.3 (and its "analog" for Model II, i.e., Assertion 10.11) directly follows from the well-known sufficient conditions of Nash equilibrium existence in the continuous games, see [64].

The example below admits a unique Nash equilibrium.

Example 10.3 Choose $F(x) = x, H_\alpha(x) = x, H_\beta(x) = 1 - x,\ c_\alpha(\alpha) = -\ln(1 - \alpha)$, and $c_\beta(\beta) = -\lambda \ln(1 - \beta)$. The first-order optimality conditions yield $\beta = (1/\lambda)\alpha$. Under $\lambda = 1$, we obtain $\alpha^* = (1/4)$ and $\beta^* = (1/4)$. In this case,

$$x^{I*}(\alpha^*, \beta^*) = 1/2, f_\alpha(\alpha^*, \beta^*) = f_\beta(\alpha^*, \beta^*) \approx -0.2.$$

Interestingly, in the equilibrium both Principals have smaller values of the goal functions than in the "*status quo*" *profile* $(0; 0)$ (since $f_\alpha(0, 0) = 1$ and $f_\beta(0, 0) = 0$). Here the utilitarian solution is also the zero strategy profile.•

Model II. Consider the goal functions of Principals 1 and 2 of the form (10.1) and (10.2), respectively, except that α is replaced by δ, and β by γ.

Assertion 10.4 In Model II described by the antagonistic game with the zero control costs, there exists no finite DSE or NE in the Principals' game.

Assertion 10.4 is immediate from boundedness of the admissible strategy sets of the Principals and from monotonicity of $x^*(\delta, \gamma)$ in both variables (see Assertion 10.1). In addition, these properties guarantee the following result.

Assertion 10.5 In Model II described by the antagonistic game with the zero control costs, let the admissible strategy sets of the Principals be bounded: $\delta \leq \delta_{max}, \gamma \leq \gamma_{max}$. Then there exists the DSE $\delta^{DSE} = \delta_{max}, \gamma^{DSE} = \gamma_{max}$ in the Principals' game.

Consider the case of nonzero control costs.

Assertion 10.6 In Model II satisfying the conditions of Assertion 10.1, let $x^*(\delta, \gamma)$ be a continuous function, the payoff functions of the Principals be bounded, linear or concave in their strategies and the cost functions be convex with the zero derivatives at the zero point and infinite growth as the argument tends to infinity. Then there exists a finite Nash equilibrium in the Principals' game.

Assertion 10.6 holds true, since under its conditions the goal functions of the Principals are concave in their strategies and take nonnegative values on the bounded value set of the arguments, i.e., a Nash equilibrium exists in this continuous game by the sufficient conditions [64].

Example 10.4 Choose $F(x) = x$, $H_\delta(x) = x$, $H_\gamma(x) = 1 - x$, $c_\delta(\delta) = \delta^2$, and $c_\gamma(\gamma) = \lambda^2 \gamma^2$. According to Example 10.1, the CBE is $x^*(\delta, \gamma) = \delta/(\delta + \gamma)$. The goal functions of the Principals have the following form:

$$f_\delta(\delta, \gamma) = \delta/(\delta + \gamma) - \delta^2, \tag{10.5}$$

$$f_\gamma(\delta, \gamma) = 1 - \delta/(\delta + \gamma) - \lambda^2 \gamma^2. \tag{10.6}$$

The goal functions (10.5) and (10.6) are concave in δ and γ, respectively. The first-order optimality conditions yield the Nash equilibrium $\delta^* = \sqrt{\frac{\lambda}{2} \frac{1}{1+\lambda}}$, $\gamma^* = \frac{1}{\sqrt{2\lambda}} \frac{1}{1+\lambda}$.

In this case, the CBE is $x^*(\delta^*, \gamma^*) = \frac{\lambda}{1+\lambda}$, and in the NE the goal functions are $f_\delta(\delta^*, \gamma^*) = \frac{\lambda(1+2\lambda)}{2(1+\lambda)^2}$, $f_\gamma(\delta^*, \gamma^*) = \frac{\lambda+2}{2(1+\lambda)^2}$.

The utilitarian CUF $f(\delta, \gamma) = f_\delta(\delta, \gamma) + f_\gamma(\delta, \gamma)$ achieves its maximum (takes the unit value) on the zero strategy profile. The NE value of the utilitarian function is $f(\delta^*, \gamma^*) = 1 - \frac{\lambda}{(1+\lambda)^2}$, i.e., the quantity $\frac{\lambda}{(1+\lambda)^2}$ characterizes how "worse" is the NE value of the utilitarian CUF than its optimal value.•

10.3   Threshold Goal Functions

For practical interpretations, an important case concerns the *threshold goal functions* of the Principals, i.e.,

$$H_{\alpha(\beta)} = \begin{cases} H_{\alpha(\beta)}^+ & \text{if } x \geq (\leq) \theta_\alpha(\theta_\beta), \\ H_{\alpha(\beta)}^- & \text{otherwise,} \end{cases} \tag{10.7}$$

where $H_{\alpha(\beta)}^+ > H_{\alpha(\beta)}^-$. That is, Principal 1 obtains a higher payoff when the proportion of active agents is not smaller than a threshold $\theta_\alpha \in [0, 1]$, while Principal 2 obtains a higher payoff when the proportion of active agents exceeds a threshold $\theta_\beta \in [0, 1]$. Denote by $\hat{x}$ the CBE in the absence of the Principals' control actions: $\hat{x} = x^*(0, 0)$. Make a pair of assumptions as follows.

Assumption A.1 The attainability set W is the unit segment, $x^*(\alpha, \beta)$ is a strictly monotonic continuous function of its arguments, and the cost functions of the Principals are strictly monotonic. See the corresponding sufficient conditions above or check these conditions in each specific case.

Assumption A.2 Under the zero strategy of Principal 2, 1 can independently implement any CBE from $[\hat{x}, 1]$; under the zero strategy of Principal 1, Principal 2 can independently implement any CBE from $[0, \hat{x}]$.

The structure of the goal functions of the Principals and Assumptions A.1 and A.2 directly imply the following. For Principal 1 (Principal 2), it appears non-beneficial to implement the CBE exceeding the threshold θ_α (being strictly smaller than the threshold θ_β, respectively).

Model I. Define the Nash equilibrium (α^*, β^*):

$$\begin{cases} \forall \alpha \in [0; 1] \ \ H_\alpha(x^*(\alpha^*, \beta^*)) - c_\alpha(\alpha^*) \geq H_\alpha(x^*(\alpha, \beta^*)) - c_\alpha(\alpha), \\ \forall \beta \in [0; 1] \ \ H_\beta(x^*(\alpha^*, \beta^*)) - c_\beta(\beta^*) \geq H_\beta(x^*(\alpha^*, \beta)) - c_\beta(\beta). \end{cases}$$

First, consider the special case $\theta_\beta = \theta_\alpha = \theta$.

Introduce the notation $\alpha(\theta) = \min\{\alpha \in [0, 1] | x^*(\alpha, 0) = \theta\}$, $\quad \beta(\theta) = \min\{\beta \in [0, 1] | x^*(0, \beta) = \theta\}$.

Define the set

$$\Omega_{\alpha, \beta}(\theta) = \{(\alpha, \beta) \in [0, 1]^2 | x^*(\alpha, \beta) = \theta,$$
$$c_\alpha(\alpha) \leq H_\alpha^+ - H_\alpha^-, c_\beta(\beta) \leq H_\beta^+ - H_\beta^-\}, \tag{10.8}$$

which includes the pairs of strategies implementing the CBE θ with the following property: each Principal gains not less than by using the strategy that modifies his payoff (10.7). By analogy with [59, 74], set (10.8) will be called the *domain of compromise*.

By definition, if the domain of compromise is nonempty, then implementing the CBE θ in terms of the utilitarian CUF guarantees a payoff to the agents that is not smaller than in the case preserving the status quo profile $\hat{x}$. Moreover, the Principals obviously benefit nothing by implementing any CBE (perhaps, except $\hat{x}$ or θ).

Assertion 10.7 If $\theta_\beta = \theta_\alpha = \theta$ and Assumption A.1 holds, then there may exist NE of the two types only:

(1) $(0; 0)$ is the NE under

$$\hat{x} \leq \theta \text{ and } c_\alpha(\alpha(\theta)) \geq H_\alpha^+ - H_\alpha^- \tag{10.9}$$

or

$$\hat{x} \geq \theta \text{ and } c_\beta(\beta(\theta)) \geq H_\beta^+ - H_\beta^-; \qquad (10.10)$$

(2) the set of NE includes the set $\Omega_{\alpha,\beta}(\theta)$ if the latter is nonempty.

If, Assumption A.2 also holds, then
$(\alpha(\theta); 0)$ is the NE under

$$\hat{x} \leq \theta \text{ and } c_\alpha(\alpha(\theta)) \leq H_\alpha^+ - H_\alpha^-; \qquad (10.11)$$

$(0; \beta(\theta))$ is the NE under
$$\hat{x} \geq \theta \text{ и } c_\beta(\beta(\theta)) \leq H_\beta^+ - H_\beta^-. \qquad (10.12)$$

Now, we explore the connection between the domain of compromise and the utilitarian solution. Denote by

$$C(\theta) = \min_{(\alpha,\beta) \in \Omega_{\alpha,\beta}(\theta)} [c_\alpha(\alpha) + c_\beta(\beta)] \qquad (10.13)$$

the minimum total costs of the Principals that are incurred by implementing the CBE θ. In the case under consideration, the utilitarian solution satisfies the following conditions:

- if $\hat{x} \leq \theta$, then $f(\hat{\alpha}; \hat{\beta}) = \max\{H_\alpha^- + H_\beta^+; H_\alpha^+ + H_\beta^+ - C(\theta)]$;
- if $\hat{x} \geq \theta$, then $f(\hat{\alpha}; \hat{\beta}) = \max\{H_\alpha^- + H_\beta^+; H_\alpha^+ + H_\beta^+ - C(\theta)]$.

And so, if for $\hat{x} \leq \theta$ we have $C(\theta) \leq H_\alpha^+ - H_\alpha^-$ and, for $\hat{x} \geq \theta$, $C(\theta) \leq H_\beta^+ - H_\beta^-$, then the domain of compromise includes the utilitarian solution.

The example below demonstrates the crucial role of Assumption A.2 for the NE structure.

Example 10.5 Choose Пусть $F(x) = x, \theta = 1/2, H_\alpha^- = H_\beta^- = 0, H_\alpha^+ = H_\beta^+ = 1,$ $c_\alpha(\alpha) = -\ln(1 - \alpha)$, and $c_\beta(\beta) = -\ln(1 - \beta)$. Clearly (see Example 10.3), the zero strategy profile is not an NE. According to the results of Example 10.2 and expressions (10.8)–(10.13),

$$\Omega_{\alpha,\beta}(1/2) = \left\{ (\alpha, \beta) \in [0, 1]^2 \left| \frac{\alpha(1-\beta)}{\alpha+\beta-2\alpha\beta} = 1/2, \ln(1-\alpha) \geq -1, \ln(1-\beta) \geq -1 \right. \right\}$$

i.e., $\Omega_{\alpha,\beta}(1/2) = \{(\alpha, \beta) \in [0, 1]^2 | \alpha = \beta, 0 < \alpha, \beta \leq 1 - 1/e\}$. In this example, the ε-optimal utilitarian solution is the Principals' strategy profile $(\varepsilon, \varepsilon)$, where $\varepsilon \in (0, 1 - 1/e]$.•

Next, consider the general case when the Principals' thresholds appearing in the payoff functions (10.7) are different. In terms of applications (informational confrontation), the most important relationship between the thresholds is described by

$$\theta_\beta < \hat{x} < \theta_\alpha \tag{10.14}$$

Define the following functions:

$$C_\alpha(x, \beta) = \min_{\{\alpha \in [0;1] | x^*(\alpha,\beta)=x\}} C_\alpha(\alpha), \quad C_\beta(x, \alpha) = \min_{\{\beta \in [0;1] | x^*(\alpha,\beta)=x\}} C_\beta(\beta).$$

(whenever minimization runs on the empty set, we believe that a corresponding function is $+\infty$).

Since the cost functions are nondecreasing and the payoff functions have form (10.7), the Principals do not benefit by implementing the CBE from the interval $(\theta_\beta, \theta_\alpha)$ in comparison to the status quo profile $\hat{x}$. Introduce another assumption actually relaxing Assumption A.2.

Assumption A.3 Under the zero strategy of Principal 2, Principal 1 can independently implement the CBE θ_α; under the zero strategy of Principal 1, Principal 2 can independently implement the CBE θ_β.

The result below is immediate from the definition of Nash equilibrium and the properties of the Principals' goal functions.

Assertion 10.8 Under Assumptions A.1, A.3 and condition (10.14), the Nash equilibria in the Principals' game have the following characterization:

- $(0; 0)$ is the NE under

$$\begin{cases} H_\alpha^+ - c_\alpha(\alpha(\theta_\alpha)) \leq H_\alpha^-, \\ H_\beta^+ - c_\beta(\beta(\theta_\beta)) \leq H_\beta^-; \end{cases} \tag{10.15}$$

- $(\alpha(\theta_\alpha); 0)$ is the NE under

$$\begin{cases} H_\alpha^+ - c_\alpha(\alpha(\theta_\alpha)) \geq H_\alpha^-, \\ H_\beta^- \geq H_\beta^+ - C_\beta(\theta_\beta, \alpha(\theta_\alpha)); \end{cases} \tag{10.16}$$

- $(0; \beta(\theta_\beta))$ is the NE under

$$\begin{cases} H_\beta^+ - c_\beta(\beta(\theta_\beta)) \geq H_\beta^-, \\ H_\alpha^- \geq H_\alpha^+ - C_\alpha(\theta_\alpha, \beta(\theta_\beta)). \end{cases} \tag{10.17}$$

Model II with the threshold payoff functions of the Principals is designed by analogy to Model I: just replace α by δ, and β by γ. Let us illustrate Assertion 10.8 using an example for Model II.

Example 10.6 Choose $F(x) = 1/3 + 2x^2/3, \theta_\gamma = 0.4, \theta_\delta = 0.6, \ H_\delta^- = H_\gamma^- = 0,$ $H_\delta^+ = H_\gamma^+ = 1, c_\delta(\delta) = \delta^2,$ and $c_\gamma(\gamma) = \lambda^2\gamma^2.$ Here we calculate $\hat{x} = 1/2, \gamma(\theta_\gamma)$ $\approx 0.1,$ and $\delta(\theta_\delta) \approx 0.07.$ For $\lambda = 2,$ all conditions (10.15)–(10.17) fail and, hence, the NE does not exist.

For $\lambda = 20,$ conditions (10.15) and (10.17) fail, but condition (10.16) holds. Therefore, $(0.07; 0)$ is the NE.

Example 10.7 Within the conditions of Example 10.6, choose $\theta_\gamma = \theta_\delta = \theta = 0.4,$ and $\lambda = 20.$ Then we obtain

$$\Omega_{\delta,\gamma}(0.4) = \{\delta \in [0, 1], \ \gamma \in [0, 0.05] | \ \gamma = 0.1 + 1.5\delta\} = \emptyset.$$

The condition (10.15) takes place, i.e., $(0; 0)$ is the NE.•

If there exist no Nash equilibria, an alternative approach is to find and analyze the equilibria in secure strategies (ESS). This concept was originally suggested in the paper [50] as the equilibria in safe strategies and then restated in a simpler form (see [51, 52] for details). The ESS proceeds from the notion of a threat. There is a *threat* to a player if another player can increase his own payoff and simultaneously decrease the payoff of the given player via a unilateral deviation. An *equilibrium in secure strategies* is defined as a strategy profile with the following properties:

- all the players have no threats;
- none of the players can increase his payoff by a unilateral deviation without creating a threat to lose more than he gains.

Under Assumptions A.1 and A.2, define the following functions:

$$C_\delta(x, \gamma) = \min_{\{\delta \geq 0 | x^*(\delta,\gamma)=x\}} c_\delta(\delta), C_\gamma(x, \delta) = \min_{\{\gamma \geq 0 | x^*(\delta,\gamma)=x\}} c_\gamma(\gamma).$$

Again, if minimization runs on the empty set, we believe that a corresponding function is $+\infty.$

Using the definition of ESS (see above and the papers [51, 52]) and the properties of the Principals' goal functions, we establish the following result.

Assertion 10.9 Let Assumptions A.1 and A.2 hold in Model II. Then

(1) the equilibrium point $(\delta_{ESS}; 0)$ is the ESS if there exists a minimum nonnegative value δ_{ESS} such that

$$\begin{cases} x^*(\delta_{ESS}; 0) \geq \theta_\delta, \\ H_\delta^+ - c_\delta(\delta_{ESS}) \geq H_\delta^-, \\ H_\gamma^+ - C_\gamma(\theta_\gamma, \delta_{ESS}) \leq H_\gamma^-; \end{cases}$$

(2) the equilibrium point $(0; \gamma_{ESS})$ is the ESS if there exists a minimum nonnegative value γ_{ESS} such that

$$\begin{cases} x^*(0; \gamma_{ESS}) \leq \theta_\gamma, \\ H_\gamma^+ - c_\gamma(\gamma_{ESS}) \geq H_\gamma^-, \\ H_\delta^+ - C_\delta(\theta_\delta, \gamma_{ESS}) \leq H_\delta^-. \end{cases}$$

Example 10.8 Within the conditions of Example 10.6, choose $\lambda = 2$, which yields no Nash equilibria in the game. From the first system of inequalities in Assertion 10.9 we find that $\delta_{ESS} \approx 0.816$ implements the unit CBE. And the second system of inequalities appears infeasible, i.e., the above ESS is unique.•

Note that the choice of the thresholds in the payoff functions of the Principals and the payoffs proper can be treated as *meta control*. Really, under a known relationship between the equilibrium of the Principals' game and the above parameters, it is possible to analyze three-level models (metalevel–Principals–agents), i.e., to choose the admissible values of the parameters in the Principals' game that lead to an equilibrium implementing the desired CBE in the agents' game. We give an illustrative example.

Example 10.9 Within the conditions of Example 10.6, set $\lambda = 20$ and consider the following problem. It is required to choose the values of H_δ^+ and H_γ^+ under which the zero strategy profile becomes the NE in the Principals' game. By condition (10.15), it suffices to decrease H_δ^+ to $4.9 \cdot 10^{-4}$.

Under the same conditions, the next problem is to choose the values of H_δ^+ and H_γ^+ that implement the CBE $\theta_\gamma = 0.4$. According to the expression (10.17), it suffices to choose $H_\delta^+ \leq 0.029$ and $H_\gamma^+ \geq 4.$•

In addition to the standard normal-form games, we will study their "extensions," namely, hierarchical (Sect. 10.4) and reflexive (Sect. 10.5) games between two Principals. As a matter of fact, the examples below merely demonstrate how the corresponding classes of the game-theoretic models of informational confrontation can be described and analyzed. Their systematic treatment is the subject of further research.

10.4   Hierarchical Game

In mob control problems, the players (Principals) often make decisions sequentially. Here the essential factors are the awareness of each player at the moment of decision-making and the admissible strategy sets of the players. A certain hierarchical game can be "superstructed" over each normal-form game [74]. Moreover, it is necessary to discriminate between two settings as follows:

(1) One of the Principals chooses his strategy and then the other does so, being aware of the opponent's choice. After that, an informational impact is exerted on the agents. As a result, the distribution function of the agents' thresholds takes form (10.6) or (10.9). We will study this case below.

(2) One of the Principals chooses his strategy and exerts his informational impact on the agents. After that, the other Principal chooses his strategy and exerts his informational impact on the agents, being aware of the opponent's choice.

In Model I, both settings are equivalent, as yielding the same distribution function (10.6) of the agents' thresholds. However, they differ within the framework of Model II.

In the games Γ_1 (including the *Stackelberg games*), the admissible strategy sets of the Principals are the same as in the original normal-form game, and the Principal making the second move knows the choice of the opponent moving first. The corresponding situations can be interpreted as control and "*countercontrol*" (e.g., under a given value of α, choose β, or vice versa). If the original normal-form game allows easy analysis, yielding an explicit relationship between the equilibria and the model parameters, then the corresponding game Γ_1 is often explored without major difficulties.

Consider several examples of hierarchical games for the first setting of Model I with the threshold payoff functions of the Principals.

Example 10.10 Within the conditions of Example 10.5 with $\theta = 1/3$, Principal 1 chooses the parameter α and then Principal 2 chooses the parameter β, being aware of the opponent's choice (the so-called game $\Gamma_1(\alpha, \beta)$). It follows from expressions (10.4) and (10.10) that

$$\Omega_{\alpha,\beta}(\theta) = \left\{(\alpha, \beta) \in [0, 1]^2 | \alpha = \theta\beta/(1 - \beta - \theta + 2\beta\theta), 0 < \alpha, \beta \le 1 - 1/e\right\}.$$

If Principal 1 chooses the strategy α^S, the best response of Principal 2 has the form

$$\beta^S(\alpha^S) = \arg \max_{\beta \in [0;1]} \left[H_\beta(x^{S*}(\alpha, \beta)) - c_\beta(\beta)\right] =$$

$$\arg \max_{\beta \in [0;1]} \left[\begin{cases} 1, & \text{if } x^*(\alpha^S, \beta) \le \theta, \\ 0, & \text{otherwise}, \end{cases} + \ln(1 - \beta)\right] = \frac{2\alpha}{\alpha + 1}.$$

In other words, Principal 2 benefits from choosing the minimum value of β implementing the CBE θ under the given α^S. The goal function of Principal 1 can be rewritten as $H_\alpha(x^*(\alpha^S, \beta^S(\alpha^S))) - c_\alpha(\alpha^S) = 1 - c_\alpha(\alpha^S)$, where $0 < \alpha \le 1 - 1/e$. Therefore, the ε-optimal solution $(\alpha^{S*}, \beta^{S*})$ of the game $\Gamma_1(\alpha, \beta)$ is the pair of strategies $(\varepsilon, 2\varepsilon/(\varepsilon+1))$ yielding the Principals' payoffs $1 + \ln(1 - \varepsilon)$ and $1 + \ln(1 - 2\varepsilon/(\varepsilon+1))$, respectively. (Here ε represents an arbitrary small strictly positive quantity.) Note a couple of aspects. First, this solution is close to the utilitarian solution, since both Principals choose almost zero strategies. Second, the Principal moving second incurs higher costs.•

Example 10.11 Within the conditions of Example 10.10, Principal 2 chooses the parameter β and then Principal 1 chooses the parameter α, being aware of the opponent's choice (the so-called game $\Gamma_1(\beta, \alpha)$). It follows from expressions (10.4) and (10.10) that

$$\Omega_{\alpha,\beta}(\theta) = \left\{ (\alpha, \beta) \in [0, 1]^2 \,\big|\, \alpha = \theta\beta/(1 - \beta - \theta + 2\beta\theta), 0 < \alpha, \beta \le 1 - 1/e \right\}.$$

In this case, the ε-optimal solution of the game $\Gamma_1(\beta, \alpha)$ is the pair of strategies $\varepsilon/(2 - \varepsilon), \varepsilon$ yielding the Principals' payoffs $1 + \ln1 - \varepsilon/(2 - \varepsilon)$ and $1 + \ln(1-\varepsilon)$, respectively. Again, this solution is close to the utilitarian analog and the Principal moving second incurs higher costs.•

Based on Examples 10.10 and 10.11, we make the following hypothesis, which is well-known in theory of hierarchical games and their applications. The solutions of the games $\Gamma_1(\alpha, \beta)$ and $\Gamma_1(\beta, \alpha)$ belong to the domain of compromise (if nonempty), and the Principals compete for the first move: the Principal moving first generally compels the opponent "to agree" with a nonbeneficial equilibrium. This property appears in many control models of organizational systems (e.g., see [74]).

Now, consider the games Γ_2 where the Principal moving first possesses a richer set of admissible strategies: he chooses a relationship between his actions and the opponent's actions and then reports this relationship to the latter. Using the ideology of Germeier's theorem, one can expect the following. If the domain of compromise is nonempty, the optimal strategy of Principal 1 (first choosing the parameter α, i.e., in the game $\Gamma_2(\alpha(\cdot), \beta)$) has the form

$$\alpha^{G*}(\beta) = \begin{cases} \alpha^{S*}, & \text{if } \beta = \beta^{S*}, \\ 1, & \text{otherwise.} \end{cases} \tag{10.18}$$

In a practical interpretation, strategy (10.18) means that Principal 1 suggests the opponent to implement the solution $(\alpha^{S*}, \beta^{S*})$ of the game $\Gamma_1(\alpha, \beta)$. If Principal 2 rejects the offer, Principal 1 threatens him with the choice of the worst-case response. The game $\Gamma_2(\alpha(\cdot), \beta)$ with the strategy (10.18) leads to the same equilibrium payoffs of the Principals as the game $\Gamma_1(\alpha, \beta)$.

The game $\Gamma_2(\beta(\cdot), \alpha)$, as well as the hierarchical games for Model II are described by analogy.

10.5 Reflexive Game

It is also possible to "superstruct" *reflexive games* [75] over a normal-form game where the players possess nontrivial mutual awareness about some essential parameters. Assume that the distribution function $F(r, x)$ incorporates a parameter $r \in Y$ describing the uncertainty. Following the paper [75], denote by r_1 and r_2 the beliefs of Principals 1 and 2 about the parameter r, by r_{12} the beliefs of Principal 1 about the beliefs of Principal 2, and so on.

Example 10.12 In Model II, choose $F(r, x) = r + (1 - r)x, r \in Y = [0, 1]$, $H_\delta(x) = x, H_\gamma(x) = 1 - x, c_\delta(\delta) = \delta$, and $c_\gamma(\gamma) = \lambda\gamma$. The corresponding CBE is $x^*(\delta, \gamma) = (\delta + r)/(\delta + \gamma + r)$, yielding the Principals' goal function

$$f_\delta(\delta, \gamma) = (\delta + r)/(\delta + \gamma + r) - \delta, \tag{10.19}$$

$$f_\gamma(\delta, \gamma) = 1 - (\delta + r)/(\delta + \gamma + r) - \lambda^2\gamma. \tag{10.20}$$

If the parameter $r \in [0, 1]$ is *common knowledge* [75] between the Principals, expressions (10.19) and (10.20) give the parametric NE of the Principals' game:

$$\delta^* = \left(\frac{\lambda}{1 + \lambda^2}\right)^2 - r, \tag{10.21}$$

$$\gamma^* = \frac{1}{(1 + \lambda^2)^2} \tag{10.22}$$

And these strategies implement the CBE

$$x^*(\delta^*, \gamma^*) = \frac{\lambda^2}{1 + \lambda^2}. \tag{10.23}$$

Interestingly, the equilibrium strategy (10.22) of Principal 2 and the corresponding CBE (10.23) are independent of the parameter $r \in [0, 1]$ under common knowledge. The situation completely changes without the common knowledge about this parameter.

Let $r_1 = r_{12} = r_{121} = r_{1212} = \cdots$, i.e., Principal 1 possesses some (generally, incorrect) information r_1 about the uncertain parameter r, supposing that his beliefs are true and form common knowledge. Also, choose $r_2 = r_{21} = r_{212} = r_{2121} = \cdots = r$, i.e., Principal 2 is aware of the true value of r, considering it as common knowledge. In other words, Principal 2 does not know that the beliefs of Principal 1 possibly differ from the reality.

Using expressions (10.21) and (10.22), we calculate the *informational equilibrium* [75] of the Principals' game

$$\delta^* = \left(\frac{\lambda}{1+\lambda^2}\right)^2 - r_1, \gamma_* = \frac{1}{(1+\lambda^2)^2},$$

which implements the CBE

$$x^*(\delta_*, \gamma_*) = \frac{\lambda^2 + (r - r_1)(1+\lambda^2)^2}{1 + \lambda^2 + (r - r_1)(1+\lambda^2)^2}. \tag{10.24}$$

Clearly, in the general case the CBE depends on the awareness of the Principals and, under common knowledge ($r_1 = r$), expression (10.24) acquires form (10.23). By implementing *informational control* as meta control (e.g., by modifying the beliefs of Principal 1 about the value of the uncertain parameter), one can accordingly change the CBE.•

Example 10.13 Within the conditions of Example 10.12, Principal 2 possesses adequate awareness about the opponent's beliefs (i.e., he knows that the beliefs of Principal 1 can differ from the truth): $r_{21} = r_{212} = r_{2121} = \cdots = r_1$. Then in the informational equilibrium Principal 1 still prefers the strategy $\delta_* = \left(\frac{\lambda}{1+\lambda^2}\right)^2 - r_1$, whereas Principal 2 chooses

$$\gamma_*(r_1, r) = \frac{1}{\lambda}\sqrt{\left(\frac{\lambda}{1+\lambda^2}\right)^2 - r_1 + r + r_1 - r - \frac{\lambda^2}{(1+\lambda^2)^2}},$$

which implements the CBE

$$x^*(\delta_*, \gamma_*(r_1, r)) = \lambda \frac{\lambda^2 + (r - r_1)(1+\lambda^2)^2}{(1+\lambda^2)^2\sqrt{\left(\frac{\lambda}{1+\lambda^2}\right)^2 - r_1 + r}}.$$

Obviously, in the case of common knowledge ($r_1 = r$), we have $x^*(\delta_*, \gamma_*(r_1, r)) = x^*(\delta^*, \gamma^*)$.

Therefore, the current example shows that, in the reflexive games, the equilibria also appreciably depend on the *mutual awareness* of the players, i.e., the beliefs about the opponents' awareness, the beliefs about their beliefs, and so on [75].•

And finally, we emphasize another important aspect: the nontrivial mutual awareness of the Principals may cover not only the parameters of the distribution function of the agents' thresholds, but also the parameters of the payoff functions and/or cost functions of the Principals, and so on.

Example 10.14 Within the conditions of Example 10.12, Principal 1 possesses inadequate awareness about the parameter λ in the opponent's cost function; by-turn, Principal 2 knows the true value of this parameter, believing that Principal 1 possesses adequate awareness.

Choose $\lambda_1 = \lambda_{12} = \lambda_{121} = \lambda_{1212} = \cdots$, i.e., Principal 1 has some (generally, incorrect) information λ_1 on the uncertain parameter λ, supposing that his beliefs are true and form common knowledge. And set $\lambda_2 = \lambda_{21} = \lambda_{212} = \lambda_{2121} = \cdots = \lambda$, i.e., Principal 2 knows the true value of the parameter λ, considering it as common knowledge. Using the expressions (10.21) and (10.22), we obtain the CBE $x^* = \dfrac{\lambda_1^2}{\lambda_1^2 + \left(\frac{1+\lambda_1^2}{1+\lambda^2}\right)^2}$ implemented in the corresponding informational equilibrium. In the common knowledge case ($\lambda_1 = \lambda$), it becomes the CBE (10.23). •

Let us outline the main result of this chapter. It has been demonstrated how the stochastic model of mob control [25] (also see Chap. 5) can be supplemented by "superstructing" different game-theoretic models of interaction between the control subjects that exert informational impacts on a mob for their personal benefit. The relatively "simple" model of the controlled object (a mob) allows using the rich arsenal of game theory, namely, normal-form games, hierarchical games, reflexive games and other games.

A promising direction of future investigations lies in the identification and separation of the typical distribution functions of the agents' thresholds. This would yield the templates of control actions and solutions to informational control problems, as well as models of informational confrontation.

Chapter 11
Models of Mob Self-excitation

In the previous sections of this book, the proportion of active agents has evolved according to the difference or differential equations of type (2.7), (3.12), (7.1), (8.1), etc., with the distribution function of the agents' thresholds in their right-hand sides. If the agents' thresholds represent independent identically distributed random variables (i.e., there is a *probabilistic uncertainty* about the threshold values), then it is possible to consider the events of system exit from a given state set (i.e., the so-called *spontaneous mob excitation* or "self-excitation"), including the calculation of their probabilities. Generally, the theory of large deviations [34] is a common tool here; e.g., see the asymptotical results obtained in [27, 80]. In this chapter, we refine the theoretical large deviations-based estimates using the numerical estimates yielded by statistical testing. The derived results allow assessing the reliability of mob nonexcitability in the case when the events have extremely small probabilities for their statistical testing [81].

Model of mob behavior. Consider a finite set $N = \{1, 2, \ldots, n\}$ of agents. Each agent has some threshold $\theta_i \in [0, 1]$, $i \in N$. At step k, agent $i \in N$ chooses one of two *states* $\omega_{ik} \in \{0, 1\}$. At step $(k + 1)$, the agent's state is defined by the rule

$$\omega_{i(k+1)} = \begin{cases} 1, & \frac{1}{n}\sum_j \omega_{jk} - \theta_i \geq 0, \\ 0, & \frac{1}{n}\sum_j \omega_{jk} - \theta_i < 0. \end{cases} \quad (11.1)$$

According to rule (11.1), the agent acts if the system state $x_k = \frac{1}{n}\sum_i \omega_{ik}$ is not smaller than his threshold. The system state dynamics obeys the recurrent expression

$$x_{k+1} = F_n(x_k), \quad (11.2)$$

© Springer International Publishing AG 2017
V.V. Breer et al., *Mob Control: Models of Threshold Collective Behavior*,
Studies in Systems, Decision and Control 85,
DOI 10.1007/978-3-319-51865-7_11

where

$$F_n(x) = \frac{1}{n} \sum_{i=1}^{n} \chi(\theta_i \leq x), \tag{11.3}$$

and $\chi(\cdot)$ denotes the indicator of a set.

The papers [27, 80] considered the case with an uncertainty in the right-hand side of expression (11.1), namely, a sequence $\theta_1(\omega), \ldots, \theta_n(\omega)$ of independent identically distributed random variables with a distribution $F(\cdot)$ on a probability space $(\Omega, \mathcal{F}, \mathbb{P})$, instead of known thresholds of the agents. Such a situation occurs, e.g., when the thresholds are uncertain, but the agents are chosen randomly from the agent set with a given threshold distribution.

In this case, the empirical distribution function of the agents' thresholds has the form

$$F_n(x, \omega) = \frac{1}{n} \sum_i \chi(\theta_i(\omega) \leq x). \tag{11.4}$$

Under fixed F and x_0, the expression

$$x_1^n(\omega) = F_n(x_0, \omega),$$
$$\ldots$$
$$x_k^n(\omega) = F_n(x_{k-1}(\omega), \omega), \tag{11.5}$$
$$\ldots$$
$$x_K^n(\omega) = F_n(x_{K-1}(\omega), \omega)$$

defines a sequence of random finite sequences $\{\bar{x}^n(\omega) = (x_1^n(\omega), \ldots, x_K^n(\omega))\}$, i.e., system trajectories each obeing some distribution P_n on the space $\Re^K$ given by $P_n(A) = \mathbb{P}\{\omega : \bar{x}^n(\omega) \in A\}$. In other words, formula (11.5) describes *stochastic threshold behavior*.

For the distributions of the system trajectories, the paper [80] established the asymptotics

$$\liminf_{n \to \infty} \frac{1}{n} \log P_n(U) \geq -\inf_U H(\bar{y}) \tag{11.6}$$

for any open set $U \in \mathcal{B}(S)$, and $\limsup_{n \to \infty} \frac{1}{n} \log P_n(C) \leq -\inf_C H(\bar{y})$ for any closed set $C \in \mathcal{B}(S)$, where

$$H(\bar{y}) = \begin{cases} y_1 \ln \frac{y_1}{F(y_0)} + \sum_{k=1}^{K-1} (y_{k+1} - y_k) \ln \frac{y_{k+1} - y_k}{F(y_k) - F(y_{k-1})} + (1 - y_K) \ln \frac{1 - y_K}{1 - F(y_{K-1})}, & \bar{y} \in Y_1, \\ (1 - y_1) \ln \frac{1 - y_1}{1 - F(y_0)} + \sum_{k=1}^{K-1} (y_k - y_{k+1}) \ln \frac{y_k - y_{k+1}}{F(y_{k-1}) - F(y_k)} + y_K \ln \frac{y_K}{F(y_{K-1})}, & \bar{y} \in Y_2, \\ +\infty, & \bar{y} \notin Y_1 \cup Y_2, \end{cases} \tag{11.7}$$

with the notation $y_0 := x_0$, $0 \cdot \ln(0) := 0$, and

$$Y_1 = \left\{ \bar{y} \in [0, 1]^K : y_0 < y_1 < y_2 < \cdots < y_m = y_{m+1} \cdots = y_K \right\}, \quad m \in \{0, \dots, K\},$$
$$Y_2 = \left\{ \bar{y} \in [0, 1]^K : y_0 > y_1 > y_2 > \cdots > y_m = y_{m+1} \cdots = y_K \right\}, \quad m \in \{0, \dots, K\}.$$

Note that this asymptotics is widespread in the theory of large deviations [34].

In Chap. 3 (see expression (3.17)) we have described the identification results for the distribution functions of real online social networks. The parameter $\lambda \in (0, +\infty)$, called the "*heterogeneity coefficient*," characterizes the difference between the distribution function from the uniform one, while the parameter $\theta \in [0, 1]$ (the "*common relative threshold*") characterizes the decision process within the network.

Let us study the probability of an event $A \in \mathcal{F}$ that the random process (11.5) with a given initial condition (e.g., $x_0 = 0.2$) exceeds at some step the "exit point" $x_{exit} = 0.5$ (e.g., when the mob is excited or $\forall x \in [x_{exit}, 1] : F(x, \theta, \lambda) \geq x$, i.e., further dynamics of process (11.2) leads to the excitation of the whole mob) under different numbers n of agents in the system. Formally, this event is defined by

$$A^n = \left\{ \omega \in \Omega : \exists k \in \mathbb{N}, x_k^n(\omega) > x_{exit} \right\}. \tag{11.8}$$

In the sequel, the event A^n is called *the exit from domain* for the mob composed of n agents. The exit probability in the model depends only on the theoretical distribution function $F(\cdot)$ of the agents' thresholds and the number of agents n. For the distribution functions from the two-parameter family (3.17), denote this probability by

$$P_{exit}^n(\theta, \lambda) = \mathbb{P}\{\omega \in A^n\}, \quad F(x) = F(x, \theta, \lambda) \tag{11.9}$$

The asymptotic estimate (11.6) of the probability acquires the form

$$\lim \frac{1}{n} \log P_{exit}^n(\theta, \lambda) = - \inf_{\bar{y} \in A^n} H(F(\cdot, \theta, \lambda), \bar{y}). \tag{11.10}$$

Exit probability estimation. Estimate (11.10) can be rewritten as

$$P_{exit}^n(\theta, \lambda) = c(n, \theta, \lambda) e^{-n \inf_{\bar{y} \in A^n} H(F(\cdot, \theta, \lambda), \bar{y})}, \tag{11.11}$$

where, for all θ and λ, the value $c(\theta, \lambda)$ satisfies

$$\lim_{n \to \infty} \frac{\log c(n, \theta, \lambda)}{n} = 0. \tag{11.12}$$

And so, its rate of change is "smaller than exponentially in n." Without the "constant" $c(n, \theta, \lambda)$, calculation of the probability $P^n_{exit}(\theta, \lambda)$ with required accuracy becomes impossible. For this simple reason, the formulas of type (11.10) are also termed "*rough logarithmic asymptotics*." Asymptotics (11.10) can be applied for the numerical estimation of the probability $P^n_{exit}(\theta, \lambda)$ only under additional information about the constant $c(n, \theta, \lambda)$. In [81], this constant was approximately defined using the statistical testing-based estimate $\tilde{P}^n_{exit}(\theta, \lambda)$ illustrated by Figs. 11.1 and 11.2. For a detailed description of the statistical testing algorithm, we refer to the paper [81].

Figure 11.3 shows the contour lines of $\tilde{P}^n_{exit}(\theta, \lambda)$ defined as the functions $\lambda^n_1(\theta)$, $\lambda^n_2(\theta)$, and $\lambda^n_3(\theta)$ satisfying

$$\tilde{P}^n_{exit}\left(\theta, \lambda^n_1(\theta)\right) = 10^{-2}, \ \tilde{P}^n_{exit}\left(\theta, \lambda^n_2(\theta)\right) = 10^{-3}, \ \tilde{P}^n_{exit}\left(\theta, \lambda^n_3(\theta)\right) = 10^{-4}. \quad (11.13)$$

Using numerical optimization, the paper [81] calculated the estimate $\hat{I}(\theta, \lambda)$ of the function $I(\theta, \lambda) = \inf_{\bar{y} \in A^n} H(F(\cdot, \theta, \lambda), \bar{y})$. Next, the constant $c(n, \theta, \lambda)$ was estimated by the formula

$$\hat{c}(\theta, \lambda) = \frac{\hat{P}^n_{exit}(\theta, \lambda)}{e^{-n \inf_{\bar{y} \in A^n} H(F(\cdot, \theta, \lambda), \bar{y})}}, \quad (11.14)$$

with $n = 200$. For n making the statistical testing-based estimation of the exit probability impossible, expression (11.14) yields the exit probability estimate

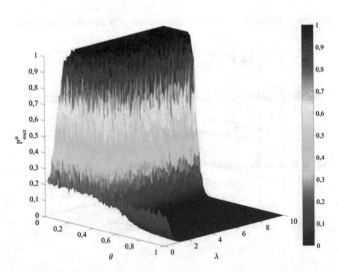

Fig. 11.1 Dependence of the exit probability $P^n_{exit}(\theta, \lambda)$ on parameters θ and λ yielded by statistical testing with n = 50

Fig. 11.2 Dependence of the exit probability $P^n_{exit}(\theta, \lambda)$ on parameters θ and λ: the graph in Fig. 11.1 projected on the plane (θ, λ)

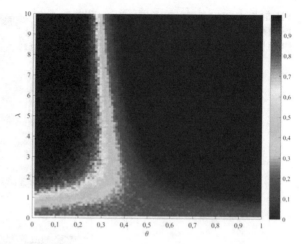

$$P^n_{exit}(\theta, \lambda) = \hat{c}(\theta, \lambda) e^{-n \inf_{\bar{y} \in A^n} H(F(\cdot, \theta, \lambda), \bar{y})}. \qquad (11.15)$$

We emphasize that this estimate actually unites the analytic result (11.10) and the numerical calculations.

Using the estimated probabilities of type (11.15) for the undesired events, one can pose and solve the corresponding control problems, e.g., choose the control parameters to minimize this probability, or minimize the costs of making this probability not greater than a given threshold, etc.

Reliability of active network structures. Consider an active network structure (a mob, a social network) with the stochastic threshold behavior (11.5). As shown in [27], this system comes to an "equilibrium" at most in n steps (i.e., the actions of the agents become invariable). Due to the randomness of the agents' thresholds, the system state corresponding to this equilibrium (the resulting state) is random.

We have earlier estimated the probability that the resulting system state leaves a given domain for different heterogeneity coefficients and common relative thresholds. The exit probability from a given domain, see above, generally depends on the system parameters.

Consider the ANS reliability problem [81], supposing that the exit probability p is a known function of the basic system parameters:

$$p = p(n, \theta, \lambda) \equiv P^n_{exit}(\theta, \lambda).$$

Let the agents' thresholds be realized (*ergo*, yielding the equilibrium) repeatedly at regular time intervals τ called "*fast time*." Then, in "*slow time*" t, the equilibrium is attained $m = [t/\tau]$ number of times, where $[\cdot]$ denotes the integer part operator.

Define *the reliability of an ANS* as the probability that its basic parameters stay within an admissible domain (for a mob, the admissible domain corresponds to its nonexcitation). For the ANS, the basic observable parameter is the mean action of

Fig. 11.3 Contour lines of exit probability for $n_1 = 50$, $n_2 = 100$, and $n_3 = 200$

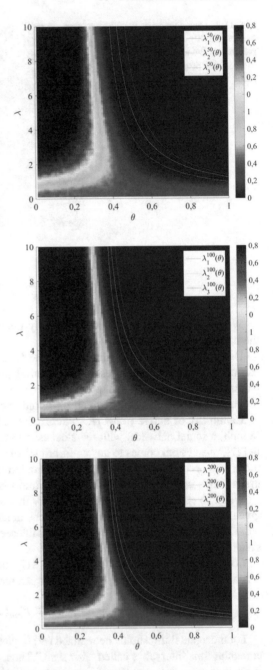

the agents (the proportion of actie agents). And the admissible domain forms the interval $[0, x_{\text{exit}})$. Since the equilibrium is attained repeatedly, the reliability of the system depends on the time interval under consideration. Designate by $R(t)$ *the survival function*, i.e., the probability that the system never leaves the admissile domain in time t. According to the definition of reliability, the survival function is the reliability of the system on the time interval t.

Introduce the random variable $\xi_i \in \{0, 1\}$ obeying the Bernoulli distribution, which is 1 if the system leaves the admissible domain at the moment $t_m = m\tau$ and 0 otherwise. The probabilities of these events are p and $(1 - p)$, respectively. The survival function can be rewritten as

$$R(t) \equiv \mathbb{P}\big(\xi_1 = \xi_2 = \cdots = \xi_{[t/\tau]} = 0\big),$$

which gives

$$R(t) = (1 - p)^{\left[\frac{t}{\tau}\right]}. \tag{11.16}$$

Under small p such that $p \ll 1/m = 1/[t/\tau]$, formula (11.16) admits the approximate representation

$$R(t) = 1 - p\left[\frac{t}{\tau}\right]. \tag{11.17}$$

The reliability control problem of the ANS is to find the set of its parameter values making the system reliability not smaller than a specified threshold δ for given time T.

Consider an illustrative example with $\delta = 0.99$ for an ANS of $n = 10^7$ agents on a time interval such that $m = \left[\frac{t}{\tau}\right] = 10^3$. The solution of this problem includes the following steps.

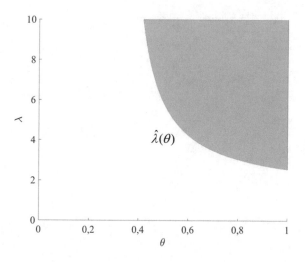

Fig. 11.4 Contour line $\hat{\lambda}(\theta)$ and parameter domain (θ, λ) making system reliability not smaller than δ

(1) Substitute the data into Eq. (11.16) and calculate the maximum admissible probability p. In the current example, we can use the approximate representation (11.17) taking the form $\delta = 1 - pm$. As a result, $p = \frac{1-\delta}{m} = \frac{10^{-2}}{10^{3}} = 10^{-5}$.

(2) Solve the equation $\hat{c}(\theta, \lambda)e^{-n \inf_{\bar{y} \in A^{n}} H(F(\cdot,\theta,\lambda),\bar{y})} = p$ and find the corresponding contour line $\hat{\lambda}(\theta)$ in the parameter space (θ, λ). The parameter domain (θ, λ) making the system reliability not smaller than δ is above this contour line, as shown by Fig. 11.4.

And finally, we underline that the reliability control problem of an ANS (keeping it within a required state set) calls for estimating the probabilities of rare events. In many cases, these probabilities are so small that their statistical testing-based estimation becomes impossible. However, the exact analytic estimates and measure replacements can be inapplicable due to the complex structure of the models. As demonstrated above and in [81], the rough logarithmic asymptotics involving large deviations can be applied for the numerical estimation of the probabilities of rare events using "calibration" of the estimate by statistical testing on the parameter set allowing such estimation.

Conclusion

The book has considered several (deterministic and stochastic, static and dynamic) models of mob control. Interestingly, there exists some "balance" between the simplicity and detailed description of agents' individual interaction that are inherent in the micromodels (on the one part) and the complexity and aggregated description of ANSs that are so characteristic to the macromodels (on the other part). Really, the ANS control problems are solved analytically either for the linear micromodel (recall the expressions (3.1), (9.1) and their analogs) or for the nonlinear macromodel (2.7). This observation is illustrated by Fig. 1.

The current and future research can be focused on the following issues that seem promising:

(1) closer interpenetration of mathematical modeling, psychology and sociology of mob;
(2) accumulation of an empirical descriptive base for ANSs, development of general identification methods;

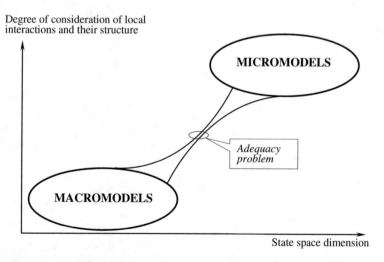

Fig. 1 Micro- and macromodels of ANSs

© Springer International Publishing AG 2017
V.V. Breer et al., *Mob Control: Models of Threshold Collective Behavior*,
Studies in Systems, Decision and Control 85,
DOI 10.1007/978-3-319-51865-7

(3) development of simulation agent-based models of mob control;

(4) development of analytic models of mob control, particularly:

- involvement of the results and methods of ANS dynamics analysis (particularly, threshold dynamics analysis) in different applications such as innovations diffusion, neural networks, genetic networks and others; see surveys in [20, 21, 48, 56, 73, 86];
- stochastic models of rare events (e.g., mob "self-excitation"), including large-deviation theory (see Chap. 11 and [27, 80, 81]);
- game-theoretic models of threshold collective behavior and their usage in a wide range of applications;
- statement and solution of control problems for the ANSs described by Granovetter's probabilistic model (8.24);
- statement and solution of dynamic informational confrontation problems for ANSs.

References

1. Akhmetzhanov A., Worden L., Dushoff J. Effects of Mixing in Threshold Models of Social Behavior. Phys. Rev. 2013. E 88, 012816.
2. Akopov A., Beklaryan L. Simulation of Human Crowd Behavior in Extreme Situations. International Journal of Pure and Applied Mathematics. 2012, Vol. 79, No. 1, P. 121–138.
3. Albert R., Barabasi A.-L. Statistical Mechanics of Complex Networks. Rev. Mod. Phys. 2002, No. 74, P. 47–97.
4. Algorithmic Game Theory (Eds. Nisan N., Roughgarden T., Tardos E., and Vazirani V.). New York: Cambridge University Press, 2009, 776 p.
5. Allport F. Social Psychology. Boston: Houghton Mifflin, 1924, 453 p.
6. Aumann R., Dreze J. Cooperative Games with Coalitional Structures. International Journal of Game Theory. 1974, Vol. 3, P. 217–237.
7. Banerjee A. A Simple Model of Herd Behavior. Quarterly Journal of Economics. 1992, Vol. 107, No. 3, P. 797–817.
8. Barabanov I., Korgin N., Novikov D., Chkhartishvili A. Dynamic Models of Informational Control in Social Networks. Automation and Remote Control. 2011, Vol. 71, No. 11, P. 2417–2426.
9. Barabanov I., Novikov D. Dynamic Models of Mob Excitation Control in Discrete Time. Automation and Remote Control. 2016, No 10.
10. Barabanov I., Novikov D. Dynamic Models of Mob Excitation Control in Continuous Time. Large-Scale Systems Control. 2016. No 63.
11. Barabasi A. Scale-free Networks. Scientific American. 2003, No. 5, P. 50–59.
12. Barabasi A., Albert R. Emergence of Scaling in Random Networks. Science. 1999, No. 286, P. 509–512.
13. Batov A., Breer V., Novikov D., Rogatkin A. Micro- and Macromodels of Social Networks. II. Identification and Simulation Experiments. Automation and Remote Control. 2016, Vol. 77, No. 2, P. 321–331.
14. Beklaryan A., Akopov A. Simulation of Agent-rescuer Behavior in Emergency Based on Modified Fuzzy Clustering / Proceedings of the 15th International Conference on Autonomous Agents and Multiagent Systems (AAMAS 2016). Singapore, 2016, P. 1275–1276.
15. Bollobas B. Random Graphs. Cambridge: Cambridge University Press, 2001, 520 p.
16. Boyd S., Parikh N., Chu E., et al. Distributed Optimization and Statistical Learning via the Alternating Direction Method of Multipliers. Foundations and Trends in Machine Learning. 2011, Vol. 3. No. (1), P. 1–122.
17. Boyd S., Vandenberghe L. Convex Optimization. Cambridge: Cambridge University Press, 2004, 727 p.
18. Breer V. A Game-theoretic Model of Non-anonymous Threshold Conformity Behavior. Automation and Remote Control. 2012, Vol. 73, No. 7, P. 1256–1264.

19. Breer V. Game-theoretic Models of Collective Conformity Behavior. Automation and Remote Control. 2012, Vol. 73, No. 10, P. 1680–1692.
20. Breer V. Models of Conformity Behavior. I. From Philosophy to Mathematical Models. Control Sciences. 2014, No. 1, P. 2–13. (in Russian).
21. Breer V. Models of Conformity Behavior. II. Mathematical Models. Control Sciences. 2014, No. 2, P. 2–17. (in Russian).
22. Breer V. Models of Tolerant Threshold Behavior (from T. Schelling to M. Granovetter). Control Sciences. 2016, No. 1, P. 11–20. (in Russian).
23. Breer V., Novikov D. Models of Mob Control. Automation and Remote Control. 2013, Vol. 74, No. 12, P. 2143–2154.
24. Breer V., Novikov D., Rogatkin A. Micro- and Macromodels of Social Networks. I. Theory Fundamentals. Automation and Remote Control. 2016, Vol. 77, No. 2, P. 313–320.
25. Breer V., Novikov D., Rogatkin A. Stochastic Models of Mob Control. Automation and Remote Control. 2016, Vol. 77, No. 5, P. 895–913.
26. Breer V., Novikov D., Rogatkin A. Models of Collective Threshold Behavior in Control Problems of Ecological-Economic Systems/Game-Theoretic Models in Mathematical Ecology. Editors: V. Mazalov, D. Novikov, G. Ougolnitsky, L. Petrosyan. New York: Nova Science Publishers, 2015, P. 1–16.
27. Breer V., Rogatkin A. Probabilistic Model of Threshold Behavior in Multiagent Systems. Automation and Remote Control. 2015, Vol. 76, No. 8, P. 1369–1386.
28. Brink R., Khmelnitskaya A., van der Laan G. An Efficient and Fair Solution for Communication Graph Games. Economic Letters. 2012, Vol. 117, P. 786–789.
29. Chen N. On the Approximability of Influence in Social Networks. SIAM J. Discrete Math. 2009, Vol. 23, P. 1400–1415.
30. Cialdini R. Influence: Theory and Practice. 5th ed. London: Pearson, 2008, 272 p.
31. Chebotarev P., Agaev R. Coordination in Multiagent Systems and Laplacian Spectra of Digraphs. Automation and Remote Control. 2009, Vol. 70, No. 3, P. 469–483.
32. Crowd Behavior at Mass Gatherings: A Literature Review/K. Zeitz, H. Tan, M. Grief, et al. Prehospital and Disaster Medicine. 2008, Vol. 24, No. 1, P. 32–38.
33. De Groot M. Reaching a Consensus. Journal of American Statistical Association. 1974, No. 69, P. 118–121.
34. Deuschel J., Stroock D. Large Deviations. New York: Academic Press, 1989, 283 p.
35. Dorogovtsev S. Lectures on Complex Networks. Oxford: Oxford University Press, 2010, 144 p.
36. Dorogovtsev S., Mendes J. Evolution of Networks. Oxford: Clarendon Press, 2010, 264 p.
37. Durett R. Random Graph Dynamics. Cambridge: Cambridge University Press, 2007, 212 p.
38. Erdos P., Renyi A. On Random Graphs. Publ. Math. Debrecen. 1959, No. 6, P. 290–297.
39. Festinger L. A Theory of Social Comparison Processes. Human Relations, 1954, No. 7, P. 117–140.
40. Forrest J., Novikov D. Modern Trends in Control Theory: Networks, Hierarchies and Interdisciplinarity. Advances in Systems Science and Application. 2012, Vol. 12, No. 3, P. 1–13.
41. Goldenberg J., Libai B., Muller E. Talk of the Network: A Complex Systems Look at the Underlying Process of Word-of-mouth. Marketing Letters. 2001, Vol. 12, No. 3, P. 211–223.
42. Gomez D., Gonzalez-Aranguena E., Manuel C., et al. Centrality and Power in Social Networks: A Game Theoretic Approach. Mathematical Social Sciences. 2003, Vol. 46, P. 27–54.
43. Goubko M., Karavaev A. Coordination of Interests in the Matrix Control Structures. Automation and Remote Control 2001. Vol. 62, No. 10, P. 1658–1672.
44. Granovetter M. Threshold Models of Collective Behavior. The American Journal of Sociology. 1978, Vol. 83, No. 6, P. 1420–1443.
45. Gubanov D., Chkhartishvili A. An Actional Model of User Influence Levels in a Social Network. Automation and Remote Control. 2015, Vol. 76, No. 7, P. 1282–1290.

46. Gubanov D., Kalashnikov A., Novikov D. Game-theoretic Models of Informational Confrontation in Social Networks. Automation and Remote Control. 2011, Vol. 72, No. 9, P. 2001–2008.

47. Gubanov D., Novikov D., Chkhartishvili A. Informational Influence and Informational Control Models in Social Networks. Automation and Remote Control. 2011, Vol. 72, No. 7, P. 1557–1567.

48. Gubanov D., Novikov D., Chkhartishvili A. Social Networks: Models of Informational Influence, Control, and Confrontation/Ed. D. Novikov Moscow: Fizmatlit, 2010, 228 p. (in Russian).

49. Idiart M., Abbott L. Propagation of Excitation in Neural Network Models. Networks. 1993, Vol. 4, P. 285–294.

50. Iskakov M. Equilibrium in Safe Strategies. Automation and Remote Control. 2005, Vol. 66, No. 3, P. 465–478.

51. Iskakov M., Iskakov A. Equilibrium Contained by Counter-threats and Complex Equilibrium in Secure Strategies. Automation and Remote Control. 2016, Vol. 77, No. 3, P. 495–509.

52. Iskakov M., Iskakov A. Equilibrium in Secure Strategies/CORE Discussion Paper 2012/61. Louvain-la-Neuve: CORE, 2012, 38 p.

53. Kempe D., Kleinberg J., Tardos E. Maximizing the Spread of Influence through a Social Network. Proc. 9th ACM SIGKDD Int. Conf. on Knowledge Discovery and Data Mining, 2003 P. 137–146.

54. Kirman A. Ants, Rationality and Recruitment. The Quarterly Journal of Economics. 1993, Vol. 108, No. 1, P. 137–156.

55. Kulivets S. Modeling of Conflict Situations with Uncoordinated Beliefs of the Agents Involving Games Based on Linear Cognitive Maps. Automation and Remote Control. 2011, Vol. 72, No. 7, P. 1568–1577.

56. Kuznetsov O. Complex Networks and Activity Spreading. Automation and Remote Control. 2015, Vol. 76, No. 12, P. 2091–2109.

57. Le Bon G. The Crowd: A Study of the Popular Mind. New York: Macmillan, 1896, 230 p.

58. Lin Y., Shi X., Wei Y. On Computing PageRank via Lumping the Google Matrix. Journal of Computational and Applied Mathematics. 2009, Vol. 224, No. 2, P. 702–708.

59. Mechanism Design and Management: Mathematical Methods for Smart Organizations/Ed. by Prof. D. Novikov New York: Nova Science Publishers, 2013, 163 p.

60. Miller D. Introduction to Collective Behavior and Collective Action Illinois: Waveland Press, 2013, 592 p.

61. Moscovici S. The Age of the Crowd: A Historical Treatise on Mass Psychology. Cambridge, Cambridge University Press, 1985, 418 p.

62. Moulin H. Cooperative Microeconomics: A Game-Theoretic Introduction. Princeton: Princeton University Press. 1995, 440 p.

63. Myers D. Social Psychology. 12th ed. Columbus: Mcgraw-Hill, 2012, 768 p.

64. Myerson R. Game Theory: Analysis of Conflict. Cambridge, Massachusetts, London: Harvard University Press, 2001, 600 p.

65. Myerson R. Graphs and Cooperation in Games. Mathematics of Operations Research. – 1977, Vol. 2, P. 225–229.

66. Nemhauser G., Wolsey L., Fisher M. An Analysis of the Approximations for Maximizing Submodular Set Functions. Mathematical Programming. 1978. Vol. 14, P. 265–294.

67. Novikov D. Big Data and Big Control. Advances in Systems Studies and Applications. 2015, Vol. 15, No. 1, P. 21–36.

68. Novikov D. Cognitive Games: A Linear Impulse Model. Automation and Remote Control. 2010, Vol. 71, No. 10, P. 718–730.

69. Novikov D. Cybernetics: From Past to Future. Berlin: Springer, 2016, 107 p.

70. Novikov D. Hierarchical Models of Warfare. Automation and Remote Control. 2013. Vol. 74, No. 10, P. 1733–1752.

71. Novikov D. Mathematical Models of Teams Building and Functioning. Moscow: Fizmatlit, 2008, 184 p. (in Russian).
72. Novikov D. Models of Informational Warfare in Mob Control. Automation and Remote Control. 2016, Vol. 77, No. 7, P. 1259–1274.
73. Novikov D. Models of Network Excitation Control. Procedia Computer Science. 2014, Vol. 31, P. 184–192.
74. Novikov D. Theory of Control in Organizations. New York: Nova Science Publishers, 2013. 341 p.
75. Novikov D., Chkhartishvili A. Reflexion and Control: Mathematical Models. London: CRC Press, 2014, 298 p.
76. Novikov D. Games and Networks. Automation and Remote Control. 2014, Vol. 75, No. 6, P. 1145–1154.
77. Owen G. Values of Games with a priori Unions/Henn R, Moeschlin O. (Eds.) Essays in Mathematical Economics and Game Theory. Berlin: Springer-Verlag, 1977, P. 76–88.
78. Ren W., Yongcan C. Distributed Coordination of Multi-agent Networks. London: Springer, 2011, 307 p.
79. Rogatkin A. Granovetter Model in Continuous Time. Large-Scale Systems Control. 2016, No. 60, P. 139–160. (in Russian).
80. Rogatkin A. Large Deviations in Social Systems with Threshold Conformity Behavior. Automation and Remote Control. 2016. (in press).
81. Rogatkin A. Probability Estimation for Rare Events in Mob Behavior. Large-Scale Systems Control. 2016. (in Russian).
82. Schelling T. Micromotives and Macrobehaviour. New York, London: Norton & Co Ltd, 1978, 256 p.
83. Schelling T. The Strategy of Conflict. Cambridge: Harvard University Press, 1960, 328 p.
84. Shibutani T. Society and Personality. Piscataway: Transaction Publishers, 1991, 648 p.
85. Shoham Y., Leyton-Brown K. Multiagent Systems: Algorithmic, Game-Theoretic, and Logical Foundations. New York: Cambridge University Press, 2008, 532 p.
86. Slovokhotov Yu. Physics and Sociophysics. I–III. Control Sciences 2012, No. 1, P. 2–20; No. 2, P. 2–31; No. 3, P. 2–34. (in Russian).
87. Zimbardo P., Leippe M. Psychology of Attitude Change and Social Influence. Columbus: McGraw-Hill, 1991, 431 p.

Printed in the United States
By Bookmasters